THE SECRETS OF BUILDING
AN ALCOHOL PRODUCING STILL

Written and illustrated by
Vincent R. Gingery

Published by
David J. Gingery Publishing LLC
P.O. Box 318
Rogersville, MO 65742

Web: http://www.gingerybooks.com

Email: gingery@gingerybooks.com

PRINTED IN THE U.S.A.

2012 REVISED EDITION

LIBRARY OF CONGRESS
CATALOG CARD NUMBER 94-76940

INTERNATIONAL STANDARD
BOOK NUMBER 1-878087-16-9

TABLE OF CONTENTS

Warning!

Although producing alcohol for fuel use is legal, *production of alcohol for human consumption is not legal.* Serious penalties can result if you are caught producing alcohol illegally.

You face serious danger both in the construction of the still mentioned in this book and in the production of alcohol from it. The author of this book is not an engineer or scientist and no engineering has been applied to this project. You should evaluate each phase of this project and proceed using your own good judgement and extreme caution. The author assumes no liability. *You are fully responsible for your own actions and the safety of those around you!* Alcohol is an extremely volatile and flammable liquid. In the right conditions even a spark from static electricity can ignite alcohol creating a horrible explosion causing property damage and serious injury. As with gasoline, the fumes not the liquid, burns. Consequently, all distilling operations must take place outdoors away from buildings and allowing for plenty of ventilation. Alcohol must be located far from any open flame or other source of ignition.

With careful attention to safety and by following all laws regarding the production of alcohol we can encourage further experimentation. This will result in the further relaxation of the laws governing the production of alcohol allowing us to become even more involved in perfecting the production of this amazing, clean burning fuel.

Preface

Having made my own wine, beer and even soda pop for the last several years I have become fascinated by the entire process involving the action of yeast on sugar. I guess that it is only natural that I would progress to the point of actually wanting to distill the alcohol from the beer and/or wine product.

Until a few years ago it was illegal to produce distilled grain alcohol for any reason. Anyone who wanted to do so had to go through mountains of red tape. An expensive bond was required to cover any possible tax liability. As far as distilling was concerned it was even illegal to own a water distillery without a permit. I believe that for this reason it is difficult even now to find detailed information on distilling alcohol.

Due to the fuel shortages in the 1970's the government decided to loosen the restraints on the production of grain alcohol, otherwise known as ethanol and in the old days known as moonshine. The desire was to encourage the development of alternative fuel sources. It is now possible to obtain an experimental permit to produce alcohol within a matter of days. The Federal permit is free and it is not necessary to post a bond.

There may, however, be some requirements at the State level.

When I decided to build an alcohol producing still, I knew very little about the distilling subject. In trying to develop the project I searched for and found only a few books to help me. Of the books I did find, none gave me detailed construction plans for a practical alcohol producing still of a size suitable for an experimenter like myself. I also found it difficult to find information on producing the mash suitable for producing grain alcohol. Because of the problems and unnecessary expense that I suffered, I decided that it might be a good idea to detail the material and methods that I used to produce ethanol from start to finish. The book you hold in your hand is the result.

The purpose of this book is to hopefully make the experimental alcohol production process easier and less expensive for those of you who might decide to attempt it. Often through experimentation at the grass roots level, major improvements and discoveries are made. I hope this book proves useful in promoting and improving the production of one of our fascinating natural resources.

Introduction

In the next few paragraphs I will give a brief description of the steps we will take in producing alcohol and an overview of the still design.

To begin with, we will discuss applying for and getting your experimental alcohol producers permit from the Bureau of Alcohol, Tobacco and Firearms. This is a simple process and it only takes a few days to get a permit.

The second step of the alcohol producing process is called malting. The process includes sprouting and drying grain. Once a grain is sprouted and then dried it is said to be malted. The sprouting stage of the grain converts nonfermentable starch to fermentable sugar. Sprouting also produces enzymes in the grain which further aid in the starch to sugar conversion process. More about this later. There are two ways to get the necessary malt required for the conversion process. The easiest way would be to buy your malted grain from someone who sells wine and beer supplies. If you buy malted barley, make sure it has all of the necessary enzymes needed for conversion. Some specialty malts do not contain them.

The other option would be to make your own malt. This is really a very simple and inexpensive process, but it is time consuming. The grain I used for making my malt was corn. The reason I use corn is because it is easier to find at the feed store. Although barley is the preferred, corn works just fine.

If you choose to make your own malt the first stage of the malting process is sprouting the grain. To do this we must create an acceptable environment for the seeds. Later on in the book I will show you how to build a small 10 layer sprouting bin to make the sprouting process easier. Each layer of the sprouting bin consists of a simple rectangular box made of 2x2 lumber with a screen bottom for drainage. Each sprouting bin holds 1 pound of grain.

The second stage would be the drying process and it takes about four days. Three ways of drying your grain will be discussed. After the grain has dried, it will be necessary to separate the sprouts from the malted grain. Once cleaned and separated, the malt is stored in a dry place until it is needed. Just before it is needed, it is ground to a fine meal to further expose the starch granules thus making it easier to turn them into sugar.

The third step is mashing. In this step all remaining starches are converted to usable sugar. Yeast can only ferment sugar. This is accomplished by mixing a measured amount of water, malted grain, and in this case ground corn together. The mixture is heated in stages. Heat is held at each stage for a specific length of time and at a specific temperature.

When all of the starch in the grain is converted to sugar the mixture is strained through a strainer bag. After the mash is strained there will be some sugar left on the grain. To get all of the available sugar from the grain we pour hot water over it. This is referred to as rinsing. The liquid that remains is called beer, or wort. The spent grains are used for livestock feed or fertilizer for the garden.

When the mashing process is complete the temperature of the wort is allowed to drop to 80 degrees at which time a yeast starter is added. The yeast starter solution is made a day ahead of time and consists of sugar, yeast, juice and a small amount of wort. The fermenting process takes about a week to ten days.

The theory of distillation; You may refer to the detailed drawing on page 10 as we discuss the operation and construction of the still.

We already know that alcohol is produced by yeast enzymes attacking the sugar in our mash solution. After all of the starch in the mash has been converted to sugar and all of the foreign particles have been strained out, it is called wort. When the wort has been fermented (all the sugar in it has turned to alcohol) it is called the wash.

The purpose of the still is to separate the alcohol from the wash. (the wash is the alcohol containing liquid remaining after fermentation is complete). The alcohol is separated from the wash by heating it to its boiling temperature.

The reason we can distill alcohol is because it mixes well with water, but the two do not form an *Azeotrope*. An azeotrope is formed when two different substances mixed together have the same boiling temperature. Water boils at 212 degrees and alcohol boils at 173 degrees. The differences in the boiling temperatures of the two are what makes alcohol distillation possible.

It should be said here that many substances that have different boiling points eventually form azeotropes. Alcohol is one of these. The critical point for alcohol is when the solution is almost pure. The azeotropic concentration for ethanol and water is 97% or 194 proof. Therefore even though our goal is to produce pure alcohol we cannot. The best performance you can expect from this still design is 170 to 190 proof alcohol, depending on how close you control the column temperature.

Your wash will probably consist of about 90% water and 10% alcohol. As the wash reaches 190 to 195 degrees the alcohol begins to rapidly separate from the water and rise up in the form of alcohol steam. Since alcohol and water combine so well there will be a certain amount of water vapor contained in the alcohol steam.

It will be necessary to separate the water vapor from the alcohol vapor before it reaches the top of the column. We are able accomplish this by packing the rectifying column full of glass marbles. As the steam rises through the column the cooler temperature of the marbles causes the water in the vapor to condense on them. The water drips off the marbles and falls back into the still tank. As the marbles warm to 173 degrees (the vaporization point of alcohol) the alcohol begins to evaporate and rise rapidly to the top of the column in its vapor form.

It is interesting to watch the changing boiling temperature of the wash as compared to the constant 173-degree column temperature. As the alcohol evaporates from the wash the boiling temperature of it gradually increases from about 190 degrees to 212 degrees. At 212 degrees there will be no alcohol left in the wash.

It is very important not to let the temperature in the upper column of the still get much hotter than 173 degrees, for this determines how pure our alcohol is. Our goal is to produce as pure an alcohol as possible. At temperatures above 173 degrees water vapor will be mixed in with the alcohol vapors (not good). Hotter temperatures in the upper column (those approaching the 200-degree range) could also cause more vapor to enter the condenser than it can safely handle. This could cause a pressure build up in the still causing the tank to rupture.

Now that alcohol vapor exists in the

upper column we need to cause it to change back into a liquid. In order for alcohol vapor to condense completely to a liquid it needs to be brought down to a temperature of 60 degrees. This is accomplished by a cooling coil. It is a length of copper tubing in the shape of a coil. It makes a spiral descent through a condenser tank. The temperature in the condenser tank is kept at 60 degrees or below by adding water to the bottom of the condenser tank and allowing it to circulate through the tank and exit out the top. By the time the alcohol vapor makes its trip through the coil it has converted to a liquid. The liquid collects in a container outside the condenser. The container is called the alcohol receiver. Now that we know how the still works we can discuss the design and how were going to build it.

The still design; At first glance the construction of the still may look difficult, but it really is not. All of the soldering is done with a hand held propane torch. Since we will not be drinking the alcohol produced from this still it will not be necessary to use the more expensive lead free solder. Lead contaminates alcohol and makes it dangerous to drink.

Since the wash that goes in the still tank is corrosive, there are only a couple of material choices that can be considered for its construction. These choices are stainless steel, or copper. Stainless is preferred, but copper is easier to work with so I chose it. The copper used comes in 3'x8' sheets and is called 16 oz. copper. It has a thickness of about .022. Since the outer tank shell is only exposed to water it can be made from 22 gauge galvanized steel. The material used in this project is easily formed and no exotic equipment is needed to build the still. All you'll need are common hand tools. I'll even show you how to make a couple of the

tools that are needed.

The outer and inner tanks are made first. The inner tank will hold a little over six gallons of wash and the outer tank will hold about twelve gallons of water.

A drain faucet is installed in each tank. Two 3/4" holes are drilled in the top of the outer tank. A 3/4" CxM adapter is soldered in each hole. CxM means copper on one end and male thread on the other end. These openings serve as fill holes and vent holes. The tank is built out of light material and

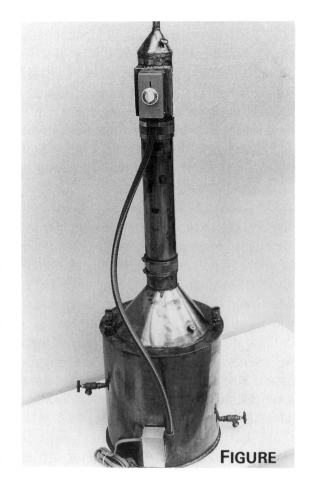

FIGURE

8

only soldered together, therefore it cannot be considered a pressure tank. It is absolutely necessary to vent the water tank as shown. Without these vents, steam pressure would build up inside the tank and cause a horrible explosion.

The water in the outer tank is heated by a 1500 watt 120 volt water heater element. When the outer tank is being heated, the wash in the inner tank is also heated indirectly. The reason the wash tank is not heated directly by the element is because the element could not withstand the corrosive qualities of the wash.

A cone serves as the still tank top. The cone has a 4-1/4" opening at the top and a short 3" section of 4-1/4" diameter copper pipe soldered to it. This opening is used to clean the still and also to add the wash to the inner tank. A thermometer is installed in the top of the cone so that it extends into the top of the inner tank. In this way we can monitor the temperature in the tank.

The 4-1/4" diam. rectifying column is built next. It is 18" long and is filled with glass marbles. The purpose of the marbles is to cause any water that is present in the vapors, as they rise from the still, to be recondensed on the marbles enabling only alcohol vapors to reach the top of the column. The column is attached to the still tank with a 4" no hub coupling. This enables the column to be easily removed so that the still tank can be charged or cleaned as the case may be.

The column top is a 10" length of 4-1/4" diam. copper pipe with a cone soldered to the top. It is attached to the rectifying column with a 4" no hub coupling. This enables the column top to be easily removed so that it and the marbles inside can be cleaned as necessary.

A thermometer is installed in the top

cone so that the upper column temperature can be monitored. An infinite range control switch is also attached to the top of the column. By monitoring the temperature and adjusting the rotary dial on the switch, the temperature of the vapors inside the column can be maintained at 173 degrees, which is the boiling point of ethanol. (An infinite range switch is the same type of device used to control the burner temperature on a standard electric kitchen stove).

A 3/4" 90 degree copper elbow is soldered to the column top. A 3/4" CxM adapter is soldered to the 90 degree elbow and an 18" long 3/4" F.I.P. x 3/4" F.I.P. water heater connecter is attached to it.

The cooling coil is a length of 5/8" type L (heavy wall) soft copper water line formed into a 3' long coil. It is then installed in the condenser tank which is a 3-1/2' section of 8" duct pipe. The duct pipe seam is soldered in order to make it leak proof. A bottom and top are made for and soldered to the condenser tank.

When in operation, water is constantly circulating through the condenser so that the alcohol vapors can be cooled from 173 degrees to 60 degrees. This cooling causes the alcohol vapors to condense to a liquid. The liquid comes out the bottom of the condenser coil as alcohol and enters the receiving tank. There is a vent located in the coil line as it exits the condenser. The receiving tank also needs to be vented or a vapor lock could occur.

As you proceed, Remember, the project is not nearly as complicated as it looks. But there are dangers involved, so proceed cautiously, use common sense and think each step through carefully.

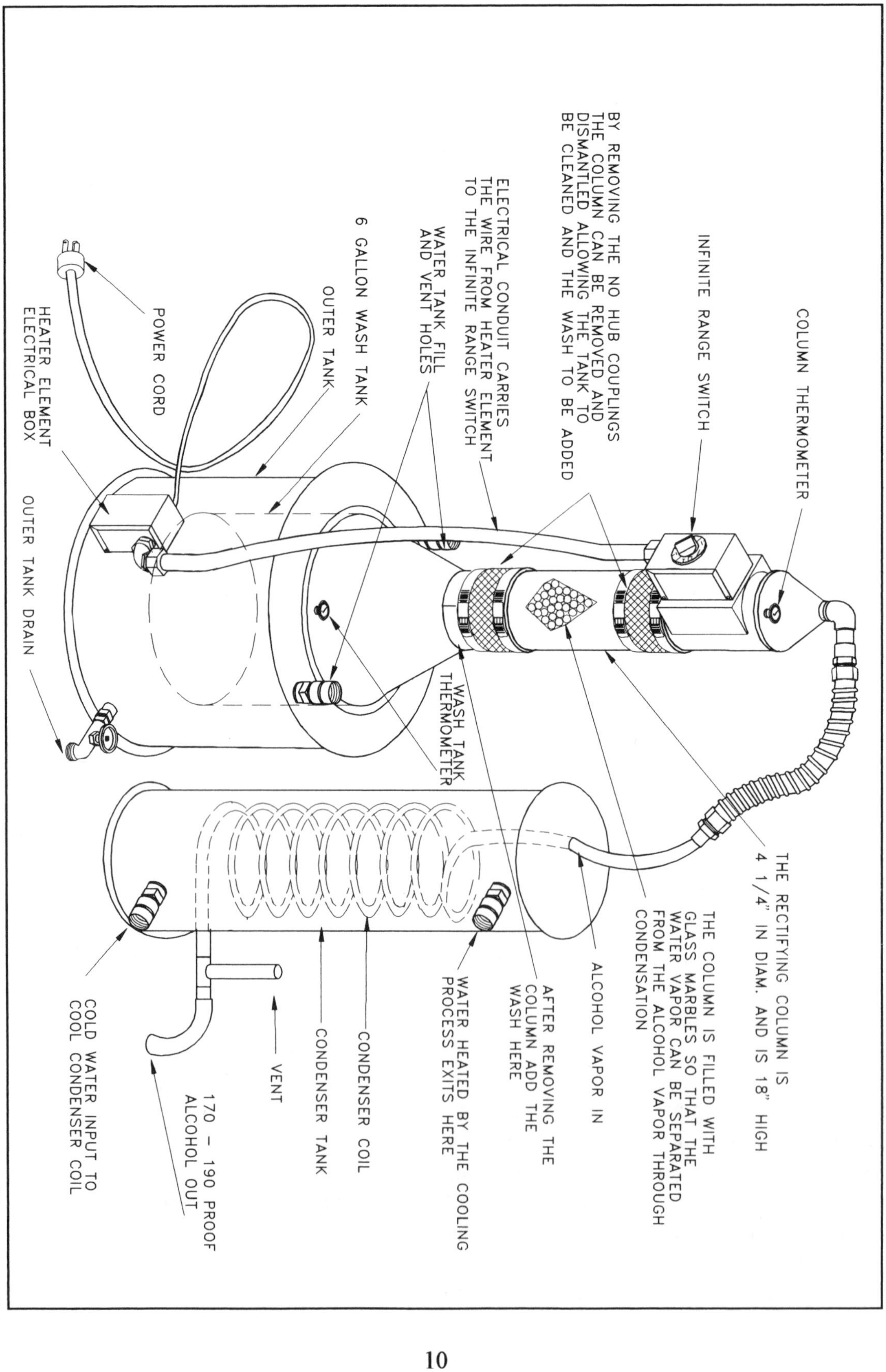

BY REMOVING THE NO HUB COUPLINGS
THE COLUMN CAN BE REMOVED AND
DISMANTLED ALLOWING THE TANK TO
BE CLEANED AND THE WASH TO BE ADDED

INFINITE RANGE SWITCH

COLUMN THERMOMETER

ELECTRICAL CONDUIT CARRIES
THE WIRE FROM HEATER ELEMENT
TO THE INFINITE RANGE SWITCH

WATER TANK FILL
AND VENT HOLES

6 GALLON WASH TANK

OUTER TANK

POWER CORD

HEATER ELEMENT
ELECTRICAL BOX

OUTER TANK DRAIN

WASH TANK
THERMOMETER

THE RECTIFYING COLUMN IS
4 1/4" IN DIAM. AND IS 18" HIGH

THE COLUMN IS FILLED WITH
GLASS MARBLES SO THAT THE
WATER VAPOR CAN BE SEPARATED
FROM THE ALCOHOL VAPOR THROUGH
CONDENSATION

ALCOHOL VAPOR IN

AFTER REMOVING THE
COLUMN ADD THE
WASH HERE

WATER HEATED BY THE COOLING
PROCESS EXITS HERE

CONDENSER COIL

CONDENSER TANK

VENT

170 – 190 PROOF
ALCOHOL OUT

COLD WATER INPUT TO
COOL CONDENSER COIL

FIGURE 2

10

Applying for your experimental alcohol producers permit

If you have decided that you want to make Ethyl Alcohol you must apply for an experimental alcohol producer permit from the Bureau of Alcohol Tobacco and Firearms. This is a simple and speedy process that only requires you to fill out a single form. As long as you are a small producer (under 10,000 gals. per year) There is no charge for the permit and no bond is required. Don't even consider not doing it. The laws governing the production of alcohol for fuel use are reasonable and the people at the ATF are friendly, courteous and helpful. Disobeying the laws and not following proper procedure is simply not worth the risk. The penalties for law breakers are serious and could result in stiff fines, jail time, or even property confiscation.

Federal laws require that every producer of ethyl alcohol, whether for beverage, industrial, or fuel use, properly qualify the plant and obtain a permit from the ATF before beginning operation.

Portions of the Crude Oil Windfall Profit Tax Act of 1980, require that a permit be obtained for producing, processing, storing, and using or distributing alcohol exclusively for fuel. The Act directs the Secretary of the Treasury, through the ATF, to expedite all permit applications, minimize any bond requirements and encourage and promote alcohol production for fuel use. Recent legislation has made the red tape involved in producing alcohol even easier. It only took three days for me to get my permit.

To obtain further information and a permit application, visit the 'Alcohol and Tobacco Tax and Trade Bureau' web site at http://www.ttb.gov/.

The information required on the permit application is minimal. There are only a few questions asked such as, a description of the still you're going to use, its capacity and the type of materials that will be used to produce the alcohol. You will also be required to submit a sketch or diagram of the plant premises. A description of security measures (such as locks, fences, building alarms etc.) to protect the premises, stills, and buildings where spirits are stored is also required.

When the application is completed, mail it to the ATF and within a few days you should receive your permit, along with a letter explaining any other necessary instructions.

You must keep records of alcohol produced in accordance with part 19, Sections 982 through 987 of title 27, code of Federal Regulations. The records must be kept at the plant where the operation occurs and must be available for inspection by any ATF officer during business hours.

You must also file an annual report of your operations for each calendar year ending December 31. (ATF Form 5110.775). The report is due by January 30 following the end of the calendar year.

Alcohol that is to be removed from the plant premises, such as for fuel in an automobile to be operated on public roads, must be rendered unfit for beverage use in accordance with a formula prescribed in 27 CFR 19.992. However alcohol used for fuel on the premises of your fuel plant and alcohol transferred to other plants need not be rendered unfit for beverage use. Alcohol must be stored under lock and key and access limited to personnel essential to the operation.

Before you begin your production check with the State and local authorities in your area. Some require licensing and bonding. If you are unsure of where to check the State regulations, the fuel tax division nearest you is a good place to start. You might also check with your State Liquor Control Division, and your State Attorney General's office.

Alcohol is a volatile, flammable, colorless liquid of a penetrating odor and burning taste. There are several different types of alcohol and they are all members of the atomic group -OH. Some of the alcohols are Methanol, Ethanol and Butanol. Ethanol, also called ethyl alcohol or grain alcohol is the alcohol contained in the beverages that we drink. Ethanol is the only alcohol that can be consumed safely (within limits). Some of the other alcohols such as methanol are highly poisonous.

The type of alcohol that we will be producing in this book is Ethanol. It is produced by a family of small microscopic plants that exist all around us and have since the beginning of time. These plants are called yeast and there are several different varieties, some good, some bad. Yeast contains enzymes. These enzymes attack the sugar and starches that exist in plant life. Yeast action on sugar is called fermentation. One primary result of fermentation is the by-product Ethanol. Another valuable by-product of fermentation is carbon dioxide which is given off as a gas and bubbles up through the liquid. Carbon dioxide is used to carbonate beverages and in the solid form, is known as dry ice. For every 100 pounds of molasses sugar fermented, approximately 46 pounds of alcohol and 44 pounds of carbon dioxide are produced. There are other by-products produced in small amounts during the fermentation process. Some of these are Propanol, Butanol, Pentanols and other higher alcohols, but we need not concern ourselves with them in this project.

Historical background; Alcoholic
beverages such as beer and wine have been with us as far back as history has been recorded. No doubt the first discovery of wine thousands of years ago was probably just an accident. Some poor guy probably took a couple of big swigs of what he thought was grape juice. Before he realized that it didn't taste quite right it was to late. Back in those days our ancestors couldn't speak English, so he looked at his wife and instead of saying "ugh" he made a face and for the first time in the human race the word "yuck" was uttered. His wife was concerned because she had made the grape juice and just couldn't understand what the problem was. She took a couple of swigs and had basically the same reaction as her husband. But by that time the warm pleasant feeling that we know so well had begun to creep over them and even though it had tasted horrible they just had to have another swig.

It is said, that this was the moment of decline for the human race. One thing for sure, on that day new words were uttered in the human language, the man and his wife had more fun than ever before and the most amazing thing was, they couldn't remember a single thing that happened. I guess that is why none of this is contained in any official historical record. Some even compare the event to Adam and Eve and the forbidden fruit, except the only immediate punishment here was a hangover. No doubt, compared to today's standards the early beverage did have a lot to be desired and the word "yuck" was probably appropriate. On a few occasions I have produced wine and beer that would compare with the taste of that early beverage.

It was only natural that improvements would be made in the original beverage, but people in those days went about things in a slow way and nothing much was done until the Greeks and Romans began making significant improvements in the process around 200 BC. Beer and wine were still the

13

drink of the day but the quality was much better. The liquid was being drained and strained. Herbs were being added for flavor and beverages were stored in sealed containers.

In certain areas of the world the water was so bad the people drank wine instead. There is even a verse in the Bible that suggests that a person drink wine for their stomachs sake. As I have shown, the history of the fermented beverage is somewhat vague and shrouded in myth. Since little was understood about the process it was considered magical.

Even though it is supposed that Aristotle was familiar with distillation three or four hundred years before Christ and that the Chinese were distilling rice beer in 800 BC, the actual discovery of the distillation of alcohol did not come into great prominence until about the 12th century AD. The primitive distilled wine of the age was hailed by 12th century alchemists as the water of life. During this period these individuals were looked upon as something more than human, and the liquid they produced from wine was thought to have unlimited curative powers. They called this liquid the spirits of the wine and employed it as the primary medicinal treatment of the age. The late 13th century Spanish philosopher Raymond Lully described alcohol as, "an element newly revealed to man but hidden from antiquity because the human race was then too young to need this beverage destined to revive the energies of modern decrepitude." (I wonder if the people that made those claims had ever suffered a hangover.)

The first stills were primitive and dangerous to say the least. The alcohol produced from them was weak and could be compared to a mild 30% brandy. By running

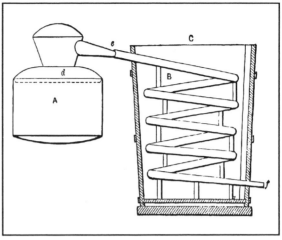

FIGURE 3 This is a drawing of a simple pot still, taken from the book "Distillation of Alcohol and Denaturing" by "F. B. Wright". An excellent book that contains lots of information about distillation and still design, available from "Lindsay Publications".

A, is the still tank which is made of tinned copper. C, is the condenser, it could have been made from metal or wood. B, is the coil, made of tinned copper pipe.

The liquor is boiled at A, often by an open fire. The vapors pass off into the worm B, which is surrounded by cold water in the condenser. The distillate is drawn off at f.

The boiler is made in two parts; The upper part fitting the lower part at d. This enables the operator to thoroughly cleanse the boiler before putting in a new lot of liquor. The joints at e, and d were sealed with dough formed by mixing flour with a small portion of salt and moistening with water. This packing prevents the escape of steam and vapor.

the wash through twice they were only able to produce a 40-50% product.

One such still was called a pot still which was heated by an open fire with the fuel being wood, coal, or whatever. Pot stills are still used to this day to produce some of the finest wine brandies made, two of those being, the French Cognac and Armagnac. These brandies are distilled in crook neck pot stills, the methods having

14

hardly changed since the early 17th century.

The quality of distillation was determined by the ability of the operator to control the temperature of the fire. If the fire became too hot it caused not only the alcohol vapors to rise but also the water vapors, creating more bulk than the worm or cooling coil could carry off. This caused pressure to build up in the still with the result being an explosion. The experienced operator would beat the connecting pipe between the still and the worm with an iron rod. See fig. 3. If a hollow sound was emitted that meant that everything was all right. If a dull sound was produced that meant that the still was running foul (clogged up) and that an explosion was eminent. In this case every effort was made to put out the fire and cool the still until the worm became clear.

In Ireland and Scotland whiskey was being made in the 12th century from malted barley, dried in a kiln over a peat fire. It was then diluted with water, cooked and left to ferment. It was later distilled using old-fashioned pot stills. Virtually every country has its specific type of alcoholic beverage and a history behind it.

The method of measuring alcohol by proof began in England. A mixture of alcohol and water was applied to gun powder. The amount that would just allow ignition of the powder was considered to be 100 proof. The amount established by this method was 11 parts ethanol to 10 parts of water. In the United States the proof figure is twice the percentage of alcohol by volume. For example, a liquor that is 90 proof contains 45 percent ethanol.

It was the Scottish and Irish settlers in Pennsylvania who first distilled whiskey in the United States. It became a successful farm industry during the period, primarily because it was easier to produce and sell than the grain from which it was derived.

The first tax placed on alcohol in the U.S. was enacted by Congress in 1791. This caused a great uproar in 1794 and resulted in what is known as the whiskey rebellion. Many distillers preferred to move to the uncharted west and face Indians rather than tax collectors. Alcohol has been taxed in the U.S. since then, except for a period from 1815 to 1862. When the Civil War began it caused a need for revenue and the liquor tax was reinstated. Since then the tax on distilled liquor has done nothing but go up up up! From 20 cents per proof gallon in 1862 to the present day tax of $10.50 per proof gallon.

Because of the misuse of alcohol by many and the mood of the nation, driven by such groups as the Women's Christian Temperance Movement and the Salvation Army, the 18th Amendment was passed and went into effect on January 16, 1920. The Volstead Act was also passed around this time providing for drastic enforcement of the amendment. This was the beginning of national prohibition in the United States. The prohibition period between 1920 and December 5, 1933 could probably be compared to the drug war being raged in our country today.

After the repeal of prohibition State Governments assumed the task of regulating liquor traffic. The Federal Government assumed all of the controlling functions, and its job included overseeing the licensing and inspecting of plants and firms having anything to do with the manufacture, importation and distillation of alcohol. Federal regulations control all aspects of the liquor industry.

It should be said that the Federal Governments interest here is probably not the health, welfare and well being of the

15

public. Its interests lay in the collection of tax revenue and its strict regulations are in force so that tax payments cannot be easily avoided. In the last few years the Government has relaxed its laws on alcohol production pertaining to fuel use to encourage the production of alternate fuel sources.

Naturally over the years many improvements have been made in the production of alcohol and many people have contributed. Even so, there is still room for improvement both in fermentation and in distilling methods. Scientists do not completely understand the entire process but they do have many formulas and theories to help explain it. I don't understand most of their technical jargon and I won't burden you with it in this book. That is not to say that those things are not important, it just is not necessary for us to concern ourselves with them at this time.

Malting the grain

Malting is the process of causing seed grains to sprout by soaking them with water after which time they are dried. The malting process will decrease the weight of the beginning grain by up to 20%. This means that if you start out with 10 pounds of grain you will end up with about 8 pounds of malt.

After the sprouted grain is dried it is said to be malted. This means that the starches that were present in the grain have been converted to sugar. Malting also produces enzymes in the grain. These enzymes are able to act on the starches of other unmalted grains converting them to sugar. For example you can use 8 lbs of *Adjunct* (normal dried, ground grain) and mix it with 2 pounds of a malted grain and

the enzymes present in the malt will convert all the starches present in the adjunct to sugar. Of course this process requires water and heat to be applied to the mixture, but more about that later.

Barley is the grain that is commonly malted and is available commercially in its malted form. One must be careful when purchasing barley malt because some specialty types do not contain the enzymes necessary for conversion. If you're uncertain ask your supplier. Your barley malt will also need to be ground. Some suppliers will do this for you free of charge. It might be a good idea to ask before you buy.

The other option for getting the necessary malt would be to make your own. Even though purchasing malt is much easier and probably in the long run less expensive, it doesn't hurt to know how to make your own.

Making your own malt; The grain we will use in our malting process will be whole grain feed corn. Yes that's right, the same stuff we feed the animals. It's available at your local feed store. Make sure it's feed corn, not seed corn. Seed grains are much more expensive than feed grains because they have been treated with chemicals and besides who wants to add a bunch of chemicals to their alcohol operation. As mentioned earlier, barley is the grain of choice when it comes to producing malt. It performs a lot better than other grains and it adds certain flavors to a finished product. But since we are making this alcohol for fuel, we should not be concerned about the flavor. I was unable to find barley grain in its whole form at any of the local feed stores in our area, so I used corn.

The first part of the malting process is sprouting. To successfully carry out the operation it is necessary to create an acceptable environment for the seeds. It might also be a good idea to take this opportunity to learn more about the sprouting process. It's really amazing.

Most of us are probably familiar with bean and alfalfa sprouts which are common ingredients in salads and some oriental food. What isn't generally known, is that sprouts are as nutritious as meat, and as rich in vitamin C as fruit. The vitamin content of the seed is increased dramatically with sprouting. Vitamin C can increase as much as 500% and B2 1300%. Folic acid, niacin, and riboflavin also increase by large percentages. Vitamins A, E, and K also go up in varying amounts. In simple terms, fats and starches in the seed are converted into vitamins, simple sugars and proteins as they absorb water and air. There are also several enzymes created in the process that promote the conversion of starch into sugar.

Seeds are divided into two basic parts. The embryo and the endosperm. The embryo is a miniature replica of the plant. The endosperm is the supply of stored carbohydrates, oils and proteins. The seed sprouts when there is warmth, moisture and air. When sprouting the embryo begins to feed upon the endosperm. At this time the inactive endosperm is turned into the nutrients mentioned above. Our desire is to harvest the malted grain before the sprouts use all the nutrients up in the growth process. The time of harvest is when the sprout has reached a length of about 1/2". There will also be three or four roots 1 to 1-1/2" long growing from each seed.

Getting started; Begin by purchasing a bag of whole grain feed corn. It generally comes in 50 pound bags and is available at your local feed store. The seeds will need to be soaked for 24 hours before they are placed in the sprouting trays. This softens them a little and it is also a good time to get rid of the imperfect seeds and foreign articles present in the grain. You can tell which is good and which is bad because the bad floats to the top.

The next step will be building the sprouting trays. If you keep your sprouting operation inside the house as I did, you will need to set your sprouting trays in or on some kind of tray or container to catch the surplus water run off. The size of the sprouting trays is not critical. I simply sized mine to fit the 28 qt. 23"x16"x6" plastic Rubbermaid box that I happened to have around the house. With that in mind we can look at the material list.

Sprouting tray material list;

Some type of plastic container (to set sprouting trays in). The container I used was a Rubbermaid 28 qt. clear box measuring 23 x 16 x 6.

Seven- 8' long 1 x 2 pine boards (for constructing tray frames).

Twenty pcs. of 3/4 x 1/8 wood strips 17-1/2" long (Screen molding.)

Twenty pcs. of 3/4 x 1/8 wood strips 13" long. (Screen molding.)

One piece of 13 x 19 x 3/4 plywood (lid).

16 feet of 36" wide standard window screen wire (for tray bottoms)

A little bit of wood glue, a few finishing nails and some paint.

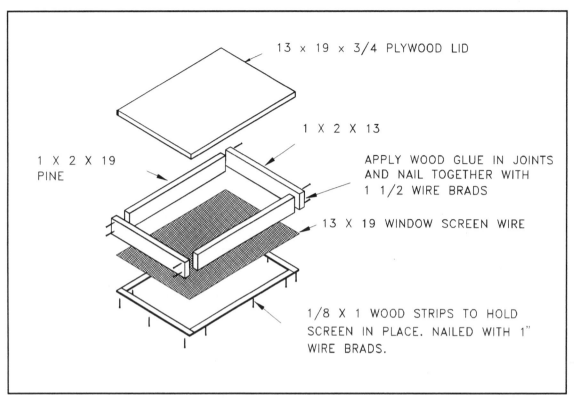

13 x 19 x 3/4 PLYWOOD LID

1 X 2 X 13

1 X 2 X 19
PINE

APPLY WOOD GLUE IN JOINTS
AND NAIL TOGETHER WITH
1 1/2 WIRE BRADS

13 X 19 WINDOW SCREEN WIRE

1/8 X 1 WOOD STRIPS TO HOLD
SCREEN IN PLACE. NAILED WITH 1"
WIRE BRADS.

FIGURE 4   Sprout tray construction

The drawing above will show you how the sprouting trays are built. The material list on page 17 includes enough material to build ten sprouting trays. An additional frame is built without a screen bottom and serves as a bottom spacer. The spacer keeps the grain in the bottom tray above the water that collects in the drain pan.

After the sprouting trays are built and painted, load each one with a portion of the corn that has been soaking. Each tray will hold about one pound of corn making it possible to convert ten pounds of regular corn to malted corn in each session. Spread the corn out evenly on the bottom of each tray to about an inch from the outside edge. The corn will be in a layer about 1-1/2" deep

toward the center of each tray. As you fill the trays stack them on the bottom spacer inside the drain pan one on top of the other. Finally, place the lid on top of the stack.

Sprinkle about a quart of water over the grain in the top tray twice a day. Because of the screen bottom, water will drip down through each tray. Once a day rotate the trays top to bottom. Also empty any water that has collected in the bottom of the drain pan. During the process you may notice mold growing on some of the grain. That's no big deal, just remove the moldy grain when you see it. The sprouting grain will develop a strong musty smell, so if you carry out the operation in the house, it might be a good idea not to invite any company

18

over while it's going on. The seeds will develop a small sprout within a day or two. Within 3 to 4 days roots will be forming. The entire process takes about seven days depending on how warm it is at the time.

The sprouts are ready for drying when they reach about 1/2 - 3/4" in length. Each sprout will also have about 3 to 4 rootlets growing from it.

There are a couple of different ways to

FIGURE 6 Drying sprouts in an American Harvest food dehydrator

FIGURE 5 Sprouts ready for drying

approach the drying process. I happened to have an American Harvest food dehydrator, so that is the method I used. I used ten dehydrator trays. Each tray holding one pound of sprouts. It took three days to dry the grain. Each day the temperature was gradually increased from 100 degrees the first day, 125 degrees on the second day and 145 degrees on the third day. There are other methods to dry your sprouts, one being the oven method. This method also takes about 3 days. Place the sprouts on large flat sheets, such as cookie sheets or pizza pans. Place on racks in the oven and follow the heating schedule below. Be sure and turn the malt over twice a day so that it drys evenly.

Day 1; 90 to 100 degrees
Day 2; 100 to 120 degrees
Day 3; 140 degrees

Another method that is not quite as desirable, but it does work in a pinch, is air drying. Simply spread the grains out thinly in a clean dry spot such as in front of an open window. A small breeze would be helpful. Leave for about 3 to 4 days turning once a day.

When the grain is dry, the brittle sprouts will have to be removed. Desprouting is a simple process. Any cloth sack will work. I simply dug up an old pillow case. Dump the grain into the bag. Keep the end tightly closed and bang the bag against a hard surface a few times. You can also rub the grain between you're hands and the dried and brittle sprouts will break

19

right off. After the sprouts have been removed place the grain in a colander, or sieve. Shake it around and the sprouts will sift out the bottom. Better not do this over a clean kitchen floor.

Now that the grain has been desprouted and dried we can call it malt. You'll notice it has a nice sweet, nutty smell. Although the malt is ground before it is used in the mash mixture it is better to store it in its whole form until needed. Ground malt tends to loose its enzyme qualities when stored while whole malt does not. Store your whole malt in a sealed container such as a plastic bucket or bag.

Gristing or grinding the malt; If you elect to purchase your own barley malt you may not have to worry about grinding your own. Some wine and beer suppliers sell malt already ground, others often have a grinder available that can be used to grind the malt you purchase.

If you are like me and make your own malt, you'll need a way to grind it. The commercial distilleries use what is called a roller mill to do all of their grinding. The grain is crushed between two rotating grooved rollers. I have seen small versions of the roller mill on the market, but they are expensive.

The next best thing and not nearly as expensive (typically $40.00-$50.00) is the hand driven flour mill. One such machine is the Corona Grain Mill. See fig. 7. It has two adjustable grooved plates that turn and crush the corn or other grain as it is forced between them. This is the method I use and it works very well.

If nothing else you could use a coffee grinder that is set on the coarse setting. This would take a little longer but probably would work in a pinch. Someone else suggested I use a rolling pin for grinding. I

tried it and it didn't work worth a hoot. So unless you're desperate I wouldn't try the rolling pin idea.

FIGURE 7 Grinding the malt in a Corona Grain Mill

Mashing;

Mashing is the process we go through to prepare the malt and *Adjunct* mixture for the fermenter. *Adjuncts* are the part of the mash mixture that is not malt. The adjunct that I use is corn because it's cheap and easy to get in my area. Some of the other unmalted cereal grains and vegetable starches you could use are, barley, oats, potato, rice, rye, sorghum, millet, tapioca, triticale and wheat. Mashing converts all of the starches that are present in the grain or vegetable matter to sugar. The sugars are then converted to alcohol through fermentation. The mashing process is carried out in three stages by combining measured amounts of ground malted grain and adjunct with water at controlled temperatures. The three stages are called *Premalt, Gelatinization, and Saccharification.*

Enzymes, and how they effect the mashing process; Enzymes are a group of proteins produced by living cells. If you remember, they were magically produced when we sprouted our grain. They have the power to initiate or accelerate specific chemical reactions in plant or animal life without themselves being destroyed. Enzymes are not living organisms, but they are programmed to react under the right conditions. They are an important tool in the mashing process and a little goes a long way. That is why most mashing recipes only call for a small amount of malt grain. Typically 20% or less of malt and 80% adjunct. There are two types of enzymes that we are concerned with. They are, *Protease* which are enzymes that digest protein and *Diastase* which are enzymes that convert starch into dextrin and sugar.

Protease are the enzymes that do the most work in the first or premalt stage of mashing, which lasts about 30 minutes. They work best at about 122 degrees. During premalt a couple of important things happen. The Protease enzymes digest protein and produce nutrients that are necessary as food for the yeast during the fermentation stage.

The other important thing that happens in the premalt stage is liquification. Without liquification the mash would be thick and impossible to work with. Because Protease and Diatase work well together the Diatase is able to convert the available starches in the malt to sugar. When the starch in the malt turns to sugar it becomes liquid. This transformation makes the mixture soupier and easier to work with and is called liquification.

Now that we have produced the necessary nutrients in the mash and it is thinned by liquification, making it easier to work with, let's discuss the second step which is gelatinization.

Starch granules are separated by protective cellulose walls. As long as the starch granules are protected, the Diatase enzymes cannot do their job. To break these walls down it is necessary to raise the temperature of the mash mixture to 212 degrees and hold it there for 15 minutes for gelatinization. Corn requires a higher temperature to gelatinize than some other grains. If you were mashing another smaller grain other than corn the higher temperatures would not be necessary for gelatinization. Generally the smaller grains gelatinize at around 155 degrees.

Now that the starch has gelatinized we are ready for the third step saccharification. In this step all of the rest of the starch present in the mixture is turned to sugar. The process takes about an hour. Temperature is critical in this stage. The Diatase enzyme works best at approximately 155 degrees. Performance is greatly reduced below 150 degrees and the action stops altogether at 170 degrees. To start the process it is necessary to allow the mash to cool to about 155-160 degrees. At that temperature the rest of the malt prescribed in the recipe is added. The mixture is stirred constantly and the temperature is monitored making sure that it does not rise above 160 degrees or fall below 150 degrees. After about 1 to 1-1/2 hours all of the starch in the mash will have been converted to sugar. Enough explanation, let's look at the recipe.

Making the yeast starter; If you plan to use a yeast starter, it's a good idea to start it about a day before you start the mashing process. To make the yeast starter put 12 ounces of orange, or grape juice in a 1 quart fruit jar. The temperature of the juice should be 80 degrees. Add a tablespoon of sugar and stir until the sugar dissolves. Add a package of wine or beer yeast to the mixture

FIGURE 8 Mash equipment, large spoon, strap with slip buckle to hold strainer bag in place, candy thermometer, 32 qt. canner with lid, strainer bag, 20 qt. stainless steel pot, 5 gal. strainer bucket with drain hose in bottom, 5 gal. bucket to catch wort as it drains

and stir well. Cover loosely and leave over night in a warm spot. The mixture will be bubbling and foaming the next morning.

When you are through making the wort cool twelve ounces of it down to 80 degrees and add it to the yeast starter along with a teaspoon of additional sugar. After setting for another twelve hours the yeast starter will be ready to pitch. (Pitch means to toss into the wort to begin fermentation.)

The recipe;

You'll need a couple of big stainless steel or enamel pots. I used a 20 qt. stainless steel pot for my rinse water, and a 32 qt. enamel canner for cooking my mash. You

will also need a large spoon for stirring, a candy thermometer (the type that clips over the side of the pan), two large plastic buckets, a nylon straining bag size 10x23, a small bottle of tincture of iodine to check for starch conversion, and yeast starter.

Ingredients:
8 pounds of ground yellow corn
2 pounds of ground malt
9.5 gallons of water

Put 2.5 gallons of water in the mash pot. Raise temperature of the water to 130 degrees.

Fill the rinse pot with 5 gallons of water and heat it to 170 degrees. This is the rinse water. (It won't be needed until later, but it's a good idea to have it ready.)

Stage 1; When the water in mash pot reaches 130 degrees, gradually add the 8 pounds of ground corn (adjunct) and eight ounces of the malt. Stir constantly to keep lumps from forming. The addition of these dry materials should have dropped the temperature in the mash pot to about 122 degrees. If not raise or lower it until it reaches 122 degrees and hold it there for 30

FIGURE 9   Premalt

minutes. This is the premalt stage.

Stage 2; Before you begin stage 2 put on a long sleeve shirt and a pair of rubber gloves to protect your skin from being burned by the hot bubbling mash that may splatter out of the pot. Also keep your face as far from the pot as you can. Take my word for it these precautions are important. I've got the burn marks to prove it.

Raise the temperature of the mash to 170 degrees very slowly, while stirring frequently to keep the bottom of the pan from burning. As the temperature of the mash increases so does the density of it. As it gets thicker and thicker it gets harder and harder to work with. (That is it tends to bubble out of the pot and burn you). When the temperature reaches 170 degrees, thin the mixture by adding 8 quarts of hot water to it. You can get the necessary hot water from the rinse pot, but be sure and replace it, because you'll need it later. This will thin the mess down so that it will be easier to work with as we raise the temperature to boiling. Slowly raise the temperature of the mash to 212 degrees. In order for it to reach the boiling temperature it will probably be necessary to keep a cover on the pot except when stirring. Hold at 212 degrees for 15 minutes. This is the gelatinization stage. (*HINT;* If you try to heat the mash too fast or do not stir it enough it will burn the bottom of your pot). Take my word for it, cleaning a pot with burnt mash all over the bottom is not fun, not to mention the fact that it smells kind of bad when it's burning. After 15 minutes of boiling the gelatinization stage is complete. Now lower the temperature of the mash to 155 degrees as rapidly as possible. A good way to achieve a rapid cool down is to set the pot in a bathtub full of cold water. (Be careful when moving the mash pot, it's hot and heavy and

FIGURE 10 Gelatinization stage ready for stage 3, Saccharification

you sure don't want to drop it). In this way it only takes about 15 minutes to cool down. When the temperature drops to about 155 degrees take the pot out of the bathtub and put it back on the stove.

Stage 3; The temperature in this stage is critical. Turn the burner back on and stabilize the temperature of the mash at 155 degrees. Add the remaining 1-1/2 pound of malt. Maintain the 155-degree temperature while stirring frequently for about 1 to 1-1/2 hours. Remember below 150 degrees and above 160 degrees the enzymes begin to quit working. That is why our temperature range is critical. This is the saccharification stage and it is complete when all of the starch has been converted to sugar.

For some reason when I use malted corn in the conversion process I am only able to convert 90%-95% of the starch in the adjunct to sugar. This is not a serious problem because any remaining starch is converted to sugar during fermentation. I mention this here because the iodine test sample for starch will not turn brown if you are using corn malt. Using corn malt, the

23

conversion process will take about an hour. You can tell conversion has taken place because the mash will have a much thinner consistency.

If you use barley malt you can expect a complete conversion within an hour to an hour and a half. The iodine test does work if barley malt is used.

Checking for starch conversion;
1. Remove a small amount of mash and place it on a plate.
2. Add one or two drops of iodine.
3. Check color

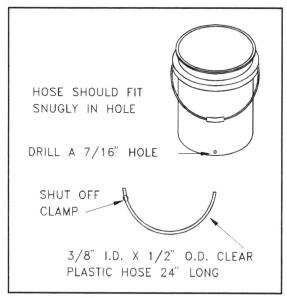

FIGURE 11 Converting a 5 gallon plastic bucket into a drain bucket.

If there is starch still present in the mash, it will react with the iodine solution to produce a bluish or purple color. If this is the case it will be necessary to carry the saccharification stage a little longer. When the starch in the mash has turned completely to sugar the iodine will remain a brown color. ***Dispose of the mash sample because it is not only toxic, but it could turn your entire batch purple or brown .***

When the conversion is complete raise

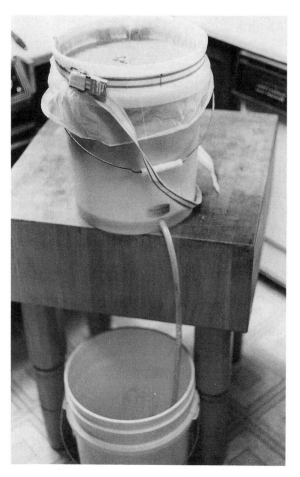

FIGURE 12 Straining the mash

the temperature of the mash to 170 degrees to stop all of the enzymatic activity. It is not necessary to hold it at that temperature for any length of time.

When the temperature reaches 170 degrees it's time to strain the mash solution.

Set the drain bucket on a sturdy table or counter top. Set the strainer bag inside the bucket and fold the ends over the edge of the bucket. Secure the bag to the bucket with a strap fitted with a slip buckle, or heavy duty clothes pins, bungee cord etc. (the strap with the slip buckle works best). Set a second five-gallon bucket on the floor beneath the drain bucket. Place the drain hose in this bucket. See fig. 12.

Dip the mash from the mash pot with a large cup and pour it slowly into the strainer bag. Pour slowly to avoid a mess. The strainer bag will stretch toward the bottom of the bucket so it may be necessary to lift it up a little higher over the edge to compensate. The strained liquid will drain out the bottom of the drain bucket, through the hose and into the receiving bucket on the floor.

Now it's time to use the 5 gallons of 170 degree rinse water that has been sitting on the stove for the last couple of hours. The purpose of rinsing is to remove any sugar that has adhered to the mash. Dip the water from the rinse pot with a large cup and slowly pour it into the strainer bag so that it runs over the spent grains. The drainage will get stuck several times so you should stir the mash in the strainer bag a little from time to time to loosen things up. Now that the liquid is strained we will refer to it as Wort.

If you have decided to use a yeast starter, cool twelve ounces of the wort to 80 degrees and add it to the starter batch at this time. Cover loosely and let it set for 12 hours.

Put a lid over the rest of the wort and allow it to cool over night.

When you're done straining you will need to dispose of the spent grains. I happen to have a few chickens so I gave it to them. They thought it was a real treat. If you have any livestock, hogs, etc. you could feed it to them. Another good place would be the garden or the compost pile.

Before beginning fermentation it's a good idea to know the potential alcohol level of the wort.

In order to find the potential alcohol level we'll have to use the hydrometer. It determines the specific gravity (SG) or density of a liquid. In this case it will tell us how much sugar we have. Since we already know that yeast converts sugar into alcohol all we have to do is find out how much sugar is in the mixture and we can make a determination about the potential alcohol level of the wort.

The more sugar there is in the wort the thicker or denser it is (the greater its gravity). The greater the gravity the better it will support anything floating in it. The hydrometer makes use of this principle. Water has been given the arbitrary specific gravity of 1.000, and other liquids are compared to this.

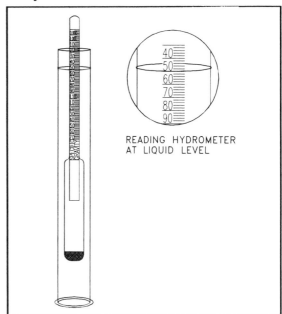

FIGURE 13   Hydrometer

Liquids that are heavier than water (in this case containing more sugar) will have specific gravities greater than water such as 1.050, 1.100, 1.130, etc.

A hydrometer is a glass tube with a weighted bulb at the bottom so it will float upright in liquid. There is a specific gravity scale as well as a potential alcohol scale printed on the side of the hydrometer.

25

The hydrometer is placed in the test jar (the container in which your hydrometer came) and a small amount of wort is added to it until the hydrometer floats. The reading is taken where the level of the liquid surface meets the scale. The thinner the liquid the deeper the hydrometer sinks. The thicker the liquid the higher the hydrometer floats. When you run a test do not return the test sample to the ferment. It could contaminate it and cause it to turn to vinegar.

Hydrometers are designed to be read at 60 degrees. If the wort is at any other temperature its reading will have to be adjusted accordingly. For every 10 degrees Fahrenheit above 60 degrees your readings will be off by .002 to .003. For example if your wort is at 80 degrees F. and it measures 1.040 you'll need to add about .004 to .006 to the reading. This would tell you that your real reading is actually about 1.045.

Using corn you can generally expect the potential alcohol reading of your wort to be between 5%-7%. Usually in the production of Ethanol alcohol the sugar produced through starch conversion is all that is used in the ferment. If one desires, additional sugar can be added to the wort to bring its potential alcohol level up to 10%, but the cost of the sugar may not justify it. One should also remember that most strains of yeast do not survive in solutions containing much more than 10%-12% alcohol. So adding too much sugar is a waste. After the potential alcohol has been checked, record the reading. You'll be able to compare that reading with the reading you take later after fermentation is complete.

The table on page 28 will show how much potential alcohol is in the ferment and how much sugar is in the gallon as compared to a particular SG reading. For

example, say you have an SG reading of 1.065. By looking across the table you can see that you have 1 lb. and 11 oz. of sugar in a gallon of wort and that the potential alcohol per gallon is 8.6%. This will not be an exact representation, because there will be solids in the wort that will affect the reading, but it will be fairly close.

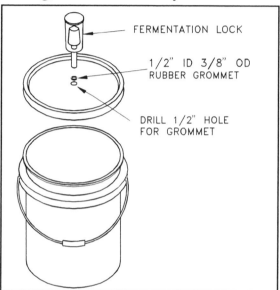

FIGURE 14 Modifying a 5 gallon bucket to serve as a fermentation bucket.

The table can also be used to tell how much sugar needs to be added to increase the potential alcohol of the wort. For example, an SG reading of 1050 shows that you have a potential alcohol reading of 6.5% and that there is 1 lb. 5 oz. of sugar in the gallon. To increase the potential alcohol level to 10% look at the chart and see what reading is closest to 10%. The closest reading is 1075, which has a 9.9% potential alcohol level. To reach this level 1 lb. 15 ounces of sugar per gallon is required. The chart shows that we already have 1 lb. 5 oz. in the gallon. To determine how much sugar is needed to increase the alcohol level, subtract 1 lb. 5 oz. from 1 lb. 15 oz. This tells us that we

26

need to add 10 oz. of sugar per gallon to increase the potential alcohol level to 10%.

Fermentation; The materials needed for the fermentation process are as follows: A 5 gal. glass jug (carboy) or a plastic closed fermenter, a fermentation lock, a plastic siphon hose, depending on which method you use to start the fermentation you'll need either a bag of yeast (any wine or beer yeast will do), or the yeast starter batch that we discussed earlier. You'll also need a hydrometer to check the potential alcohol of the wort. See Fig. 15.

Cleanliness starts here. There are many forms of bacteria that can spoil the ferment by killing the yeast or turning the alcohol in it to vinegar, so sanitation is important. Anything that will come in contact with the ferment must be sanitized. A weak household bleach and water solution of 2 cups bleach to every 5 gallons of water works well as a disinfectant. Before you begin, rinse all traces of the bleach from the ferment jug or bucket as well as the siphon hose. Any trace of bleach will kill the yeast.

After the wort has cooled to about 80 degrees and you have taken an SG reading, transfer it to a clean fermenter.

There are a couple of ways to start the fermentation. One would be to simply sprinkle a .176 oz. bag of wine or beer yeast over the top of the wort. This works fine and is certainly an easy way if you have plenty of time.

A better and faster way would be to use the yeast starter we discussed earlier on page 22.

The fermentation method we use is called a "closed fermentation." This means that the ferment is closed off from the air and environment. In this way we can be 99% assured that wort will not be contaminated by wild yeasts and other organisms.

Add the yeast to the ferment. Insert an airlock in the top opening of the fermenter. Fill the airlock half full with water. Within a few hours the mixture will begin bubbling and foaming. This process continues for about one or two weeks. After all noticeable yeast activity has stopped there will be 2 -4" of sediment on the bottom of the fermenter. If this sediment were poured in the still it would soon clog it up.

The best way to separate the liquid from the dregs (bottom sediment) is by syphoning the ferment into another container. This is referred to as racking.

After the ferment has been racked it can be left to set for a few hours, or even days

FIGURE 15 Fermentation equipment, hydrometer, siphon hose, 6 gallon plastic fermenter with air lock and 5 gallon glass carboy with air lock.

for further settling. When racking, take care not to stir in too much oxygen and other air borne material that may cause contamination.

Sometimes racking will stimulate further active fermentation that will last for as much as 2 or 3 more days. When the liquid has cleared and all visual yeast activity has ceased, check the ferment with a hydrometer to see if all of the sugar has been converted to alcohol. If you get a reading of 1.000 or less congratulate yourself. This means all of the sugar has been converted to alcohol. If all of the sugar was not converted, you can tell how much alcohol is in the mixture by comparing your hydrometer reading with the reading you took before fermentation was started.

After fermentation is complete the liquid is called the wash and it is ready for the still. The cleaner and clearer the wash the better the still will operate and the less you'll have to clean it.

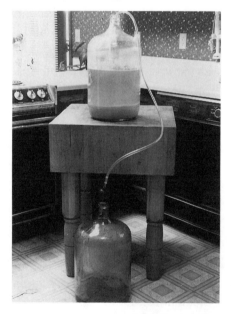

FIGURE 16   Racking the ferment

Table showing potential alcohol and amount of sugar per gallon of wash

SG	Potential % alcohol by volume	Amount of sugar in the gallon	
		lb.	oz.
1010	0.9		2
1015	1.6		4
1020	2.3		7
1025	3.0		9
1030	3.7		12
1035	4.4		15
1040	5.1	1	1
1045	5.8	1	3
1050	6.5	1	5
1055	7.2	1	7
1060	7.8	1	9
1065	8.6	1	11
1070	9.2	1	13
1075	9.9	1	15
1080	10.6	2	1
1085	11.3	2	4
1090	12.0	2	6
1095	12.7	2	8
1100	13.4	2	10
1105	14.1	2	12
1110	14.9	2	14
1115	15.6	3	0
1120	16.3	3	2
1125	17.0	3	4
1130	17.7	3	6
1135	18.4	3	8

Old ways versus new;

There are several ways to produce the beer required for alcohol production and as many different ways to distill it. The methods I have previously detailed in this book worked well for me, but it doesn't hurt to gain an insight on other methods.

The art of producing alcohol on a small scale is quickly dying to the point of being virtually nonexistent and good solid information is hard to find. Like many other things in our generation, big business is taking it over. Because of volume buying and production they can produce and sell their products at a much smaller cost thus eliminating competition.

Many of the old moonshiners considered their knowledge of alcohol production as valuable information and kept it to themselves. Each claimed to have secret methods and ingredients that made their product better than their competitors. They were not willing to share any secrets because of the fear of creating competition. Typically the moonshine business was a family tradition carried from one generation to the next.

In an effort to encourage further experimentation and to give an interesting comparison, I am including a few paragraphs on modern production methods as well as a few paragraphs explaining the old way.

Modern methods of production; As can be imagined the machinery used in large scale production is massive and the cost of putting together such an operation is beyond most of us. But it doesn't hurt to gain an understanding of how they do it.

The first step in the process consists of taking a sample of the proposed grain for analysis and inspecting it for quality. If the grain is accepted it is then unloaded by one of two methods.

The grain is usually delivered in box cars and is unloaded from chutes located under them. It is spilled into pits covered with metal grating or screens designed to remove sticks, metal, and other trash from the grain as it passes from the car to the pit. With the power shovel method a series of large adjustable shovels are located on an elevator system and used to transport the grain from the pit to the storage elevator.

Another method of unloading the grain is the Airveyor system. This system works like a large vacuum cleaner. The grain is transported by air flowing under a vacuum created by a large blower located above the grain bin. The suction end of the tube is inserted into the grain. The blower is turned on and the grain is sucked up through ductwork into a cyclone aerator where it drops out. The air used to transport the grain is vented into the atmosphere. By a series of airlocks and a proper ductwork system the Airveyor system can be used to transport grain to any desired location.

Before the grain is ground, it is cleaned to remove any foreign articles such as iron, rags, sticks, wood or other trash. Other things removed are weed seeds, such as garlic and ergot, any of which could harm the quality of the finished product.

The machine used for cleaning the grain consists of a series of vibrating screens that sift out the dirt and other foreign material. At the same time jets of air remove chaff, seeds, and dust. Dust collectors remove light material and magnetic separators remove tramp iron.

Cleaned grain is stored in storage bins located in well ventilated areas until it is needed for milling (grinding).

The grain is ground to break the outer protective walls of the kernels and to expose more surface to the action of heat and chemicals during the process of cooking, conversion and fermentation.

The most common machine used by distilleries for grinding the grain is the roller mill. The operation consists of two steps. The separation of the endosperm from the bran, and the gradual reduction of the endosperm to flour.

The first step is carried out with a series of grooved rolls that shear the kernel and gradually scrape the endosperm from the bran coat. The rolls which break the kernels are called "break" rolls and the material going to the rolls is called "break" stock.

The second step is done with smooth rolls that crush the endosperm to flour. This is known as the reduction step and the product derived from it is called "middlings." The coarse middlings are called "sizings" and the fine middlings are called "tailings."

Another method of grinding the grain is by the use of hammer mills. This method is not very popular because the texture of the ground grain cannot be controlled. The grain enters the top of the mill and is crushed by revolving hammers that rotate at speeds of 1800 to 3600 RPM.

Today's modern methods generally do not require the use of malt such as we used in our process. Instead specific enzymes are used, those being the alpha amylase enzymes used during the protein rest and the gelatinization stage and the glucoamylase enzyme used during the saccharification stage. These enzymes are produced by Miles Laboratories and are called TAKA-THERM and DIAZYME respectively.

Any one of three types of cookers may be used in the mashing process. One method is the open tub in which the mash can only be heated to 212 degrees. (This is the method we used). Another method is the batch pressure cooker. In this type of cooker the mash can be heated to any desired temperature within the design limits of the equipment. Usually a temperature of about 310 degrees is chosen. The other method is the continuous pressure cooker. The difference between it and the batch pressure cooker is that with this method the grain is exposed to sudden high temperatures for a comparatively short time.

Fermentation is taken seriously in large scale production. Yeast starters are carefully produced and all fermentation is carried out in large temperature controlled vats. An entire book could be written just on the fermentation process alone.

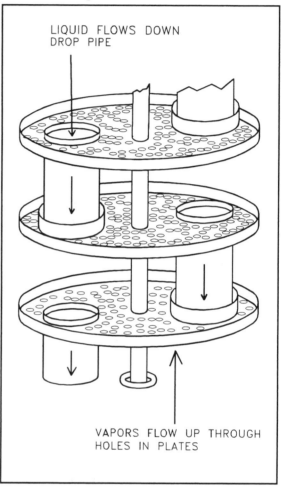

FIGURE 17   **Perforated plates from a sieve plate column**

The distillation process in an industrial alcohol plant works on the same principle as our still, except on a much larger scale. Heating is carried out by applying some means of indirect heat or by passing live steam through the fermented mash. The vapors rise up and enter a continuous fractionating column. The type most common is the sieve plate column which is divided into sections by perforated plates. This type of column operates under the same principle as ours except temperatures are more rigidly controlled and of course perforated plates are used instead of marbles.

The liquid to be distilled enters the column at a level where the temperatures are controlled to match that of the liquid on the feed plate. The plates below the feed plate are called stripping plates because they remove the alcohol from the feed. The plates above the feed plate are called rectifying plates.

In simple terms the vapor leaving one plate enters the liquid on the plate above by passing through the perforations in the plate. It condenses partially as it enters the liquid and releases heat which in turn vaporizes an equivalent amount of the liquid on this plate. The vapor passes on to the next plate and the process is repeated throughout the column. Because of this process the percentage of alcohol contained in the liquid decreases with each downward plate until the liquid is entirely free of alcohol and exits at the base of the column.

The alcohol vapor gradually rises to the top of the column and enters a tubular condenser. The vapors are condensed by cooling water which is pumped through coils that surround the condenser.

There is very little waste in a large scale alcohol operation. The carbon dioxide given off as a gas in the fermentation process is marketed to the beverage industry and used to carbonate soft drinks. It is also sold in its solid form as dry ice. The spent grains are called distillers grain and are resold as animal feed.

How the moonshiners did it; The most common product produced in the old days was called "pure corn." Sometimes sugar was used in the process to increase the alcohol yield, but this was considered cheating by many and the product could not be labeled pure corn. An experienced taster could tell the difference instantly not only by taste, but also by the bead that was produced in the liquid when the bottom of the jar containing the alcohol was hit with the palm of his hand.

Moonshining was an illegal process in the old days. In that respect nothing much has changed, because it's still illegal today.

The people that did it were often respected pillars of the community their illegal deeds being overlooked for favors, those being either cash or liquid. From what I've been able to find out this is the way they did it not so many years ago using 50 gallon barrels.

Still construction; Since the operation had to be hidden from the revenuers, a secluded location somewhere deep in the woods was generally chosen. There had to be a cold running stream close by so that the condenser coil could be cooled. Usually the still would be built downhill from the stream so a wooden slough could be constructed for that purpose.

The furnace was often constructed of clay and rocks. So if possible a location was chosen with the right soil characteristics. The furnace was built around a 50-gallon barrel that served as the still tank. A pipe was run out the bottom of the barrel so that the spent beer could be drained this being called the slop arm. A pipe also came out of the top of the still tank and went into another 50-gallon barrel. This pipe was called the cap arm. The barrel that it went to was called the thumper keg. A pipe came out the thumper keg and went to the condenser. The condenser tank could have been made of metal or wood and it had a length of copper coil inside it. Water from the creek was constantly running over the condenser coil to cool the liquor. The alcohol exits the bottom of the condenser and enters a large container.

Old timers recipe for sour mash; The information in the next few paragraphs has not been proven or tried by me. It is simply fascinating stuff that I have been able to dig up from several different sources. Health and safety were not as much of a concern in the old days. So the methods described here could be downright dangerous and the product produced could be poisonous. Use caution and ***Don't drink the stuff!***

To begin with good, fresh, white corn was preferred and it was said that it would produce three to five quarts of whiskey per bushel. Inferior hybrid or yellow corn would only produce two to four quarts per bushel. The amount of corn used in this recipe is nine bushels. Two bushels are set aside to sprout. You'll also need seven fifty-gallon barrels.

To sprout corn In the winter put it in a

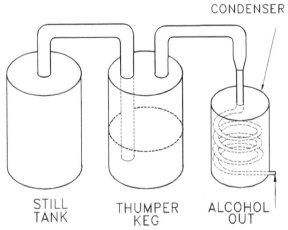

CONDENSER

STILL
TANK

THUMPER
KEG

ALCOHOL
OUT

FIGURE 18   **Barrel still**

barrel of warm water and leave it for twenty-four hours. Then drain and cover it with warm water again and leave it for fifteen minutes, then drain again. Place the tub close to a warm stove. Rotate it twice a day to distribute the warmth evenly. Each day add warm water to the tub and leave for fifteen minutes and then drain. Stir the corn once a day so that the corn on the bottom of the tub is rotated to the top. With this method it is said that you could have good malt within four or five days with shoots about two inches long and good root growth as well.

To sprout corn in the summer put it in tow sacks. Sprinkle warm water over the sacks once a day and flip them over. The sacks are sometimes covered with sawdust or manure to help hold heat. Be careful when covering the sacks because if the corn gets too hot it could go slick and not sprout. If the sprouts start getting too hot stir them up to give them air and to cool them.

About a day before the sprouted corn is ready, the remaining seven bushels of corn are taken to the miller to be ground up. Care is taken to only grind the corn, not crush it. If it is crushed it will have some heavy material in it that will sink to the bottom of the still and burn.

The fermenting operation is ready to commence. Take the ground meal to the still location in the woods. Fill the still nearly full with water, then build a fire under it. Stir in a half bushel of corn meal. Bring to a boil and let bubble for thirty to forty five minutes. When it's through cooking bring one of the empty barrels over to the still and place it under the slop arm. Pull the plug stick from the bottom of the still and let the hot mash from the still fill the barrel. Add a half bushel of uncooked meal to the barrel you just filled and let its hot contents cook it. Make sure to stir it in well. Set the barrel aside and repeat the process for the other seven bushels of corn. When you're through all seven barrels will be filled with the mash mixture.

Take the sprouted corn (malt) to the mill and have it ground. Care is taken to use a miller who can be trusted and will keep the activities secret. He wouldn't charge a fee for grinding the malt at that time, but he would take his payment later. When a miller wasn't available a sausage grinder could be used.

Back at the still, thin the mash in each barrel. This is done by placing a stick in the mash barrel so that it stands upright. Add water to the barrel until the stick no longer stands up by itself, but falls over against the side of the barrel on its own. After the mixture is thinned add a gallon of malt to each barrel and stir it in. Next sprinkle a double handful of raw rye on top of the mash in each barrel. The rye helps make a good cap, helps keep the mixture working and helps the final product hold a good bead. (If sugar is being used add 10 pounds to each barrel at this time.) Cover the barrels, because if it rains all will be ruined.

The next day the mixture in the barrels should be working. *It should be noted here that no yeast was added to the ferment in these old recipes. They relied totally on wild yeast to begin the ferment. As you can imagine, using these methods caused the odds of a successful fermentation to be slim, probably 50 - 60% or less. As you'll see our ancestors weren't so dumb. They had a way of dealing with everything. Common sense told them that if they were successful with yeast activity in one barrel and not successful in another, simply take a dipper of ferment from one of the barrels that was working and add it to a barrel that wasn't.*

In that way they were able to increase the odds of success.

Stir the mixture in each barrel once a day for next couple of days. On the fourth day if you're using sugar add a half gallon of malt and thirty-five to forty pounds of sugar to each barrel.

If sugar is not used the whole mixture will be ready to run on the fifth day. With sugar it takes about 10 days.

One could tell when the mixture was ready to run by studying the cap that forms on top of it. Sometimes the cap will be two inches thick, or sometimes a half inch thick, often it will just have suds on top. This is called a "blossom cap."

When the cap is almost gone with only a few remnants left scattered across the top the mixture is ready to run. At this point it must be run through the still immediately, or the mixture will turn to vinegar. It is better to run a day early than a day late.

Add mash to the still tank until it's three quarters full. This allows enough room for expansion due to heat. Put 10 gallons of mash in the thumper keg. (*Notice that the mash solution is not strained. Can you imagine the mess that had to be dealt with when cleaning the still?*)

Start the fire under the still tank and as it heats up stir constantly to keep the mash from sticking to the side and burning. Keep this up until the mash reaches a rolling boil.

Set the still top and cap arm in place. Seal in place and plug all cracks with a paste made of flour and salt moistened with water.

Careful attention must be given to the intensity of the fire. About fifteen minutes after the beer starts boiling in the still, the steam from it will begin to have an effect on the cold beer in the thump barrel and cause it to start bubbling and thumping making an awful racket. Usually this could be heard for several hundred yards through the woods. Don't worry though, the thumping will soon stop. When it does the beer will be boiling smoothly and you'll soon have alcohol flowing. Place a large 10 - 12 gallon container at the outlet of the condenser. A funnel is inserted in the opening to the container and is lined with a clean, fine white cloth on the bottom, a yarn cloth on top of that and a couple handfuls of washed hickory coals on top of that. The hickory coals remove the fusel oil that shows up as an oil slick on top of the liquor. If the alcohol was consumed and this were not done the fusel oil would make one very sick.

When the thumping stops the whiskey will start flowing. A couple of gushes of steam comes first. Then a strong surge of liquid that soon subsides to a trickle. Then comes the second surge which means it's coming for good. Keep running the still as long as any trace of alcohol remains.

Then drain the thump barrel and add the results of the first run to it. These results are referred to as backings and should consist of about ten gallons of weak liquor.

Drain the still through the slop arm and refill it with fresh beer.

Begin the second run. The product produced during this run will be double strength. Keep checking the alcohol content with a proof vile as it comes out of the condenser. (A proof vile is just a small glass jar with a lid.) The test is done by knocking the bottom of the proof vile against the palm of your hand and watching the bubbles that are formed. When the bubble action ceases very little alcohol remains in the liquid. Pull the container away. You should have two to three gallons of good whiskey. Put it aside and save it. Another way to check the alcohol content was to catch a small amount of the liquid and see if it would burn.

Catch the remainder of the second run in another container. This will be the backings for the third run. During the third run the liquor will be weaker because the backings in the thumper keg do not contain as much alcohol. On the fourth run you get more liquor because you will have richer backings. This cycle continues until all the mash has been run.

After you are all done you can expect to get about ten gallons of pure corn whiskey from this recipe, for an average of about one to one and a half gallon of liquor per bushel. If sugar were used the return would be about six gallons of whiskey per bushel.

The alcohol product derived from this method is called the high shots and had to be cut to be drinkable. The liquor was usually cut with water. From six gallons of high shots one could expect eight to ten gallons of fine whiskey.

It must be remembered that the production of alcohol for drinking purposes is illegal. I have included a description of the methods used years ago because some of the information has value and could be used when developing new methods of production. It is also an interesting historical comparison between the old and the new. It should also be remembered that the new could not exist without the trial and error methods of the past. So ironically the illegal actions of our ancestors helped pave the way for the modern production methods of the present.

Building The Still

Material list;

Two 3' x 8' sheets of 16 ounce copper. 16 ounce copper sheets measure about .022 in thickness. If you want to save a little money the outer tank shell can be made of 22 gauge galvanized steel. If you go that route you can substitute one sheet of galvanized steel for one sheet of copper.

One, 1 lb. spool of solder.

One small can of solder paste.

Four 1/8" pop rivets.

One, 1500 watt, 120 volt, screw in flange water heating element.

One, screw type water heater element adapter kit.

Four, 3/8-16 x 3/4" brass bolts with nuts

One, commercial 120 volt infinite range switch kit.

One, six foot long, heavy duty 15 amp, 120 volt power cord.

3-1/2' of 3/4" liquid tight conduit.

One 1/2" straight conduit connector

One 1/2" 90 degree conduit connector

Two, 5-1/4"Lx3"Wx2-1/8"D metal project boxes. (Available from Radio Shack)

Four, 3/4" FHT x 3/4" FHT adapter fittings. (FHT means female hose thread)

Two, 1/2" CxM adapter fittings. (C means copper end and M means male thread end)

Five , 3/4" CxM adapter fittings

One, 1/2" copper to 3/4" male pipe adapter.

Two, water shut off valves with 1/2" female ends.

One foot of 1/2" copper water pipe.

One foot of 1/4" copper tubing.

Two, 0-220 degree F. pocket dial thermometers

Twelve pounds of glass marbles.

Two 4" no hub couplings.

3" of 3/4" copper water line.

One, 3/4" 90 degree elbow.

One, 18" flexible copper water heater connecter, 3/4 F.I.P. x 3/4 F.I.P.

20' of 1/2" type L thickwall copper tubing.

One, 5' section of 8" galvinized furnace pipe. (Condenser tank)

Tools;

3/8" drill.

Drill bit sizes 1/8, 3/16, 1/4, 3/8, 1/2, 3/4.

One pair each of straight, right and left cutting tin snips.

One set of dividers capable of spreading at least 10".

A vise grip sheet metal tool.

A hammer with a square face such as a tinners hammer or auto body hammer.

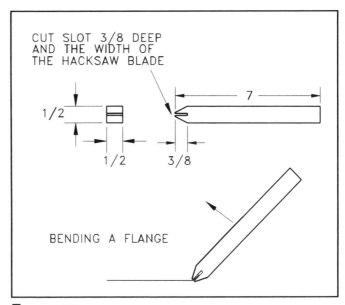

Figure 19   Making the flange bending tool

Some type of steel bar mounted to a work bench so that it hangs over the edge 18". A couple of suggestions are a 3' length of 2" diam. steel round stock, or 3' length of 2" square bar. Another suggestion would be a 3' section of old railroad track.

A piece of 1/2" round rod 12" long. (Use as a swedge tool.)

Large crescent wrench.

Assorted hand wrenches 1/4 - 5/8.

Heater element socket

Two or three 4" "C" clamps.

Bench grinder.

Assortment of screw drivers.

Propane gas torch.

Flange bending tool. This is a simple tool that we can make. It is a 7" length of 1/2" x 1/2" hot roll steel. A 3/8" deep slot the width of a hacksaw blade is cut into one end. The corners on the slot are also beveled using a bench grinder. As you will see later this tool is used to bend the flanges necessary to form the seams used for construction.

Construction methods;

Some of you may already have a good working knowledge of sheet metal construction. But for those of you who don't lets review a few of the basic skills that will be required in this project.

Up until a year ago I knew very little about the subject myself. Then my Dad wrote his latest book, "Working Sheet

36

Metal", available from Lindsay Publications. With his permission I have used many of the construction methods detailed in his book. In fact, if I had not used his book as a reference, this project would not have gone as smoothly as it did. Thanks Dad.

Safety is the first thing to consider in any shop activity. There are many hazards involved in this project both hidden and obvious and no effort has been made to point out all of them. The author of this book is not an engineer and no engineering has been applied to this project. Each detail presented is subject to your own appraisal and you are fully responsible for your own safety and that of any others who may be injured due to your pursuit of this activity.

Extra caution should be taken when working with sheet metal because its edges are sharp. Major cuts requiring stitches are common in the sheet metal trade so be careful.

Eye protection is another concern. Make it a habit to wear safety glasses in all your shop activities.

You'll also be soldering with an open flame so there will be fire hazards. All soldering must be done away from flammable materials. Be sure and watch the direction in which the flame of the torch is pointing. Always turn the flame off before setting the torch aside because it could be knocked over and cause a fire. Solder only in a well-ventilated area because the fumes

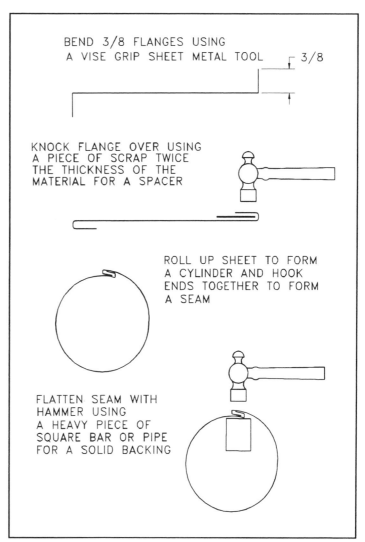

BEND 3/8 FLANGES USING A VISE GRIP SHEET METAL TOOL ⌐ 3/8

KNOCK FLANGE OVER USING A PIECE OF SCRAP TWICE THE THICKNESS OF THE MATERIAL FOR A SPACER

ROLL UP SHEET TO FORM A CYLINDER AND HOOK ENDS TOGETHER TO FORM A SEAM

FLATTEN SEAM WITH HAMMER USING A HEAVY PIECE OF SQUARE BAR OR PIPE FOR A SOLID BACKING

FIGURE 20   **Forming a folded seam**

you create are hazardous.

When drilling make sure the work is securely clamped. Also keep long hair and loose clothing out of the way when drilling.

Forming seams; The seams used in this project are basic in nature and are not difficult to form. Since we are using light gauge copper our job is fairly easy and no heavy duty equipment will be required. Light gauge copper is easily formed into neat cylinders or cones by hand.

The *folded seam* is the seam used on all of the cylinders and cones made in this project. See fig. 20. The width

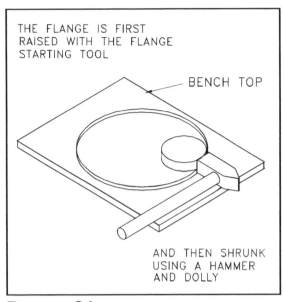

THE FLANGE IS FIRST RAISED WITH THE FLANGE STARTING TOOL

BENCH TOP

AND THEN SHRUNK USING A HAMMER AND DOLLY

FIGURE 21   Raising a flange on a disc

allowances for the seams are 3/8" on each end. These allowances have been included in each drawing.

A flange is bent on each side of the sheet metal that is to be formed. A vise grip sheet metal tool does a good job of bending the flanges. This tool is simply a pair of vise grips with 3" wide flat jaws. Another option for bending flanges would be a sheet metal brake if you're lucky enough to own one.

The object is to bend the flanges completely over to form a hook with one hook being formed in the opposite direction of the other. After bending the flanges as far as you can with the vise grip tool fold them over the rest of the way with a hammer using a piece of scrap twice the thickness of the material for a spacer. The material is then formed by hand into a cylinder or cone and the folded edges are hooked together.

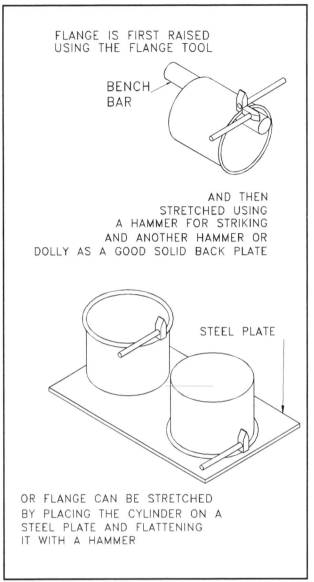

FLANGE IS FIRST RAISED USING THE FLANGE TOOL

BENCH BAR

AND THEN STRETCHED USING A HAMMER FOR STRIKING AND ANOTHER HAMMER OR DOLLY AS A GOOD SOLID BACK PLATE

STEEL PLATE

OR FLANGE CAN BE STRETCHED BY PLACING THE CYLINDER ON A STEEL PLATE AND FLATTENING IT WITH A HAMMER

FIGURE 22   Stretching a flange on a cylinder

The seam is finally formed by striking it along the hooked edges until flat with a hammer using a heavy piece of steel bar or pipe for support. A piece of pipe held in a vise works good. A square or round bar mounted to a work bench so that it extends over the edge several inches also works well. After the seam is formed solder it.

38

The single lock bucket seam; There are two elements in a single lock bucket seam and they are the cylinder and the cylinder top or bottom depending on which is being attached. The seam is used to attach the bottom or top to the cylinders thus forming a tank. The seam is formed in three steps. The steps are shown in figures 21, 22, and 23.

A 3/8" flange is formed around the disc in two steps, the first step being done with the flange bending tool described in fig. 19. A 45-degree flange is first made by making a series of partial bends with the flange bending tool. You'll notice that after the 45-degree flange is bent it will be uneven and buckled in several places. It will be necessary to straighten the flange out and bend it to 90 degrees. This is done with a hammer and dolly and the process is called

shrinking the edge.

A dolly is a heavy piece of steel used to support the blows of the hammer. In this case, since we are working on a round shape a round dolly will be required.

Set the dolly behind the flange and line up the hammer with it and gradually work your way around the disc hitting the flange with the hammer using the dolly as a back plate until it is neat and straight.

The next step will be to bend a 3/8" flange on the cylinder. Start the flange bend with the flange bending tool. As with the disc, the flange on the cylinder will be rough and uneven. To straighten it out we will have to stretch the edge. There are a couple of ways to accomplish this.

Set the cylinder on the bench bar and stretch the flange by using a dolly as a back plate and working your way around the

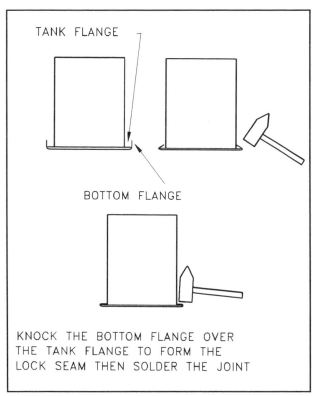

TANK FLANGE

BOTTOM FLANGE

KNOCK THE BOTTOM FLANGE OVER THE TANK FLANGE TO FORM THE LOCK SEAM THEN SOLDER THE JOINT

FIGURE 23   Forming the seam

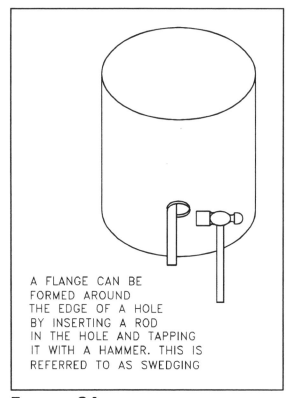

A FLANGE CAN BE FORMED AROUND THE EDGE OF A HOLE BY INSERTING A ROD IN THE HOLE AND TAPPING IT WITH A HAMMER. THIS IS REFERRED TO AS SWEDGING

FIGURE 24   Swedging a hole

flange with the hammer until it is neat and straight. See fig. 22.

The flange can also be stretched by placing the cylinder on a steel plate and flattening it with a hammer. See fig. 22.

Finally assemble the tank by placing the cylinder inside the disc. Bend the disc flange over the cylinder flange with a hammer. This forms the lock seam which is further sealed with solder. See fig.23.

Another art to learn is hole swedging which is simply forming a flange around a drilled hole. This gives us a much better fit and strengthens the joint when the fittings are soldered in.

The hole to be swedged should be drilled to a radius of 1/8" less than that required. It is then enlarged by inserting a round rod of suitable size inside the hole and tapping the side of the rod with a hammer until the hole reaches the desired size. By gradually enlarging the hole and frequently testing the fit by inserting the part that goes in it, a very good fit can be achieved. A swedged hole is also much easier to solder.

Soldering is an easy art to acquire. There are only a few important points to remember. The first thing to do is to decide what type of solder to use. My suggestion would be to use the lead free type solder. It's a little more expensive than wire solder (which contains lead). I mention this because I think the safety factor far outweighs the cost factor in instances like this. Soldering produces fumes that in turn can be breathed in and if those fumes contain lead it can only lead to harmful side effects. Remember use plenty of ventilation when soldering and be aware of the potential fire hazards as well.

The tools and materials required to solder are a gas propane torch, a friction lighter, some steel wool or emery cloth, solder, paste flux and a small brush to apply the flux with.

Before soldering be sure that the surfaces to be soldered are clean. If you are soldering shiny new copper all that will be necessary is to apply the flux and solder the joint. If however, the material is dirty or dull, clean the solder area up with steel wool or emery cloth. Then apply the flux and solder the joint.

An important point to remember is that you are heating the material to be soldered to a temperature that causes the solder to melt when it is touched to it. The temperature required to melt solder is 400-475 degrees. So heat the joint to be soldered with the torch. When the joint is hot enough the solder will flow into it by capillary attraction. Also point the flame at an angle in the direction in which you are soldering. This gives the flame a chance to heat the solder area in advance.

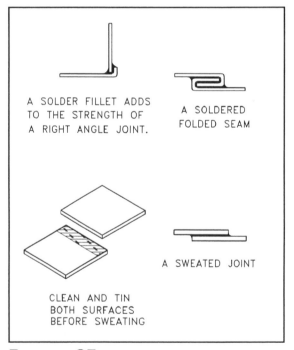

A SOLDER FILLET ADDS TO THE STRENGTH OF A RIGHT ANGLE JOINT.

A SOLDERED FOLDED SEAM

CLEAN AND TIN BOTH SURFACES BEFORE SWEATING

A SWEATED JOINT

FIGURE 25

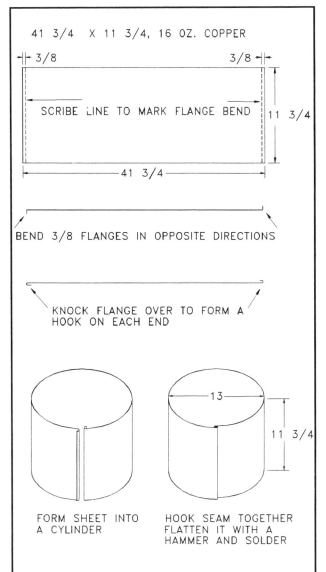

41 3/4 X 11 3/4, 16 OZ. COPPER

3/8   3/8

SCRIBE LINE TO MARK FLANGE BEND   11 3/4

41 3/4

BEND 3/8 FLANGES IN OPPOSITE DIRECTIONS

KNOCK FLANGE OVER TO FORM A
HOOK ON EACH END

13

11 3/4

FORM SHEET INTO   HOOK SEAM TOGETHER
A CYLINDER   FLATTEN IT WITH A
                       HAMMER AND SOLDER

FIGURE 26 Forming the inner tank shell

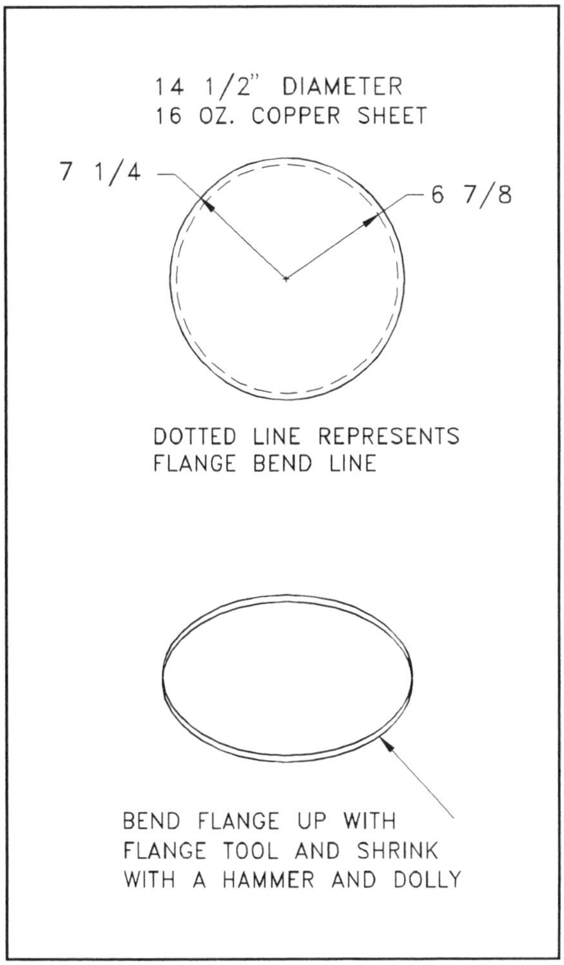

14 1/2" DIAMETER
16 OZ. COPPER SHEET

7 1/4   6 7/8

DOTTED LINE REPRESENTS
FLANGE BEND LINE

BEND FLANGE UP WITH
FLANGE TOOL AND SHRINK
WITH A HAMMER AND DOLLY

FIGURE 27   **Making a bottom
                      for the inner tank**

Step 1. Clean the joint to be soldered with steel wool, or fine, cloth backed sandpaper until bright.

Step 2. Apply a light coat of soldering flux (Soldering paste) over the area to be soldered. Special brushes can be purchased for spreading the flux or just use an old tooth brush.

Step 3. Heat the joint to soldering temperature with a torch. Do not over heat.

Test by touching with solder.

Step 4. When the joint is hot enough, solder will flow into it by capillary attraction. Flow solder into the joint until a small fillet appears around the joint. Now that we have reviewed the construction methods let's begin the project.

Building the inner tank; The inner tank cylinder is made from a sheet of 16 oz. copper that measures 41-3/4" x 11-3/4". A

41

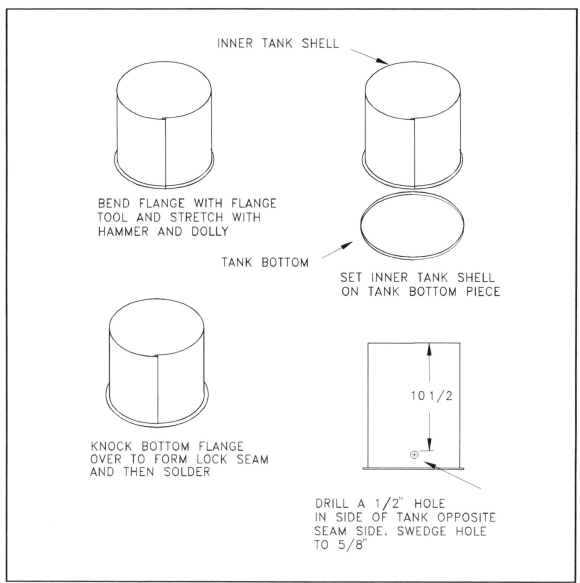

INNER TANK SHELL

BEND FLANGE WITH FLANGE
TOOL AND STRETCH WITH
HAMMER AND DOLLY

TANK BOTTOM

SET INNER TANK SHELL
ON TANK BOTTOM PIECE

KNOCK BOTTOM FLANGE
OVER TO FORM LOCK SEAM
AND THEN SOLDER

10 1/2

DRILL A 1/2" HOLE
IN SIDE OF TANK OPPOSITE
SEAM SIDE. SWEDGE HOLE
TO 5/8"

FIGURE 28   **Attaching the bottom to the inner tank cylinder**

line is scribed 3/8" from each end of the sheet showing the location of the flange bends.

Bend the flanges in opposite directions on each side of the sheet. Roll the sheet up to form a cylinder and hook the edges together to form the seam. Flatten the seam with a hammer and bench bar and then apply the flux and solder it. The finished diameter of the tank is 13". See fig. 26.

Bend a 3/8" flange on the bottom of the cylinder and stretch it with a hammer and dolly.

You'll need a set of dividers capable of spreading 10" apart to lay out the top and bottom sections of the tanks. If you do not have a set of dividers you can make a simple tool that will do the same thing. It consists

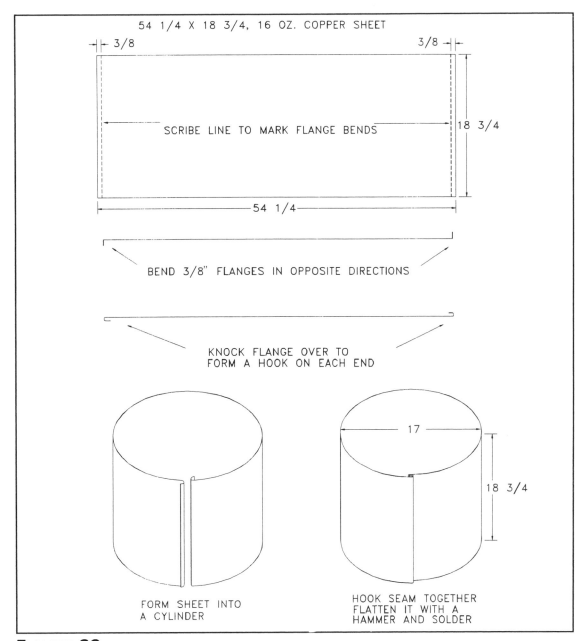

FIGURE 29   **Forming the outer tank cylinder.**

of a 1x2 board cut a couple of inches longer than the desired radius of the circle to be laid out. Hammer a sharp nail in each end of the board making the distance between the nails be that of the radius of the circle to be laid out. Put a small indention at the center point of the proposed circle with a scratch awl. Set one of the nails in the indention and scratch the circle with the other nail by moving the board in a circular direction.

The bottom of the inner tank is cut from a 16 oz. copper sheet and has a radius of 7-1/4". A 3/8" flange is bent up on its outer edge and then shrunk with a hammer and dolly. See fig. 27. Place the inner tank cylinder on the bottom section. Bend the

43

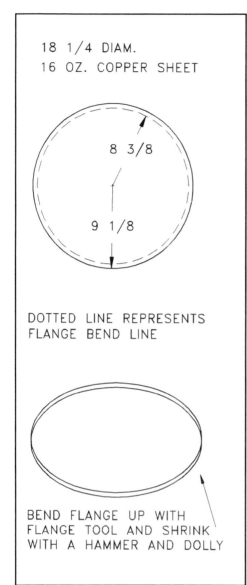

18 1/4 DIAM.
16 OZ. COPPER SHEET

8 3/8

9 1/8

DOTTED LINE REPRESENTS
FLANGE BEND LINE

BEND FLANGE UP WITH
FLANGE TOOL AND SHRINK
WITH A HAMMER AND DOLLY

FIGURE 30 Making the outer
tank bottom.

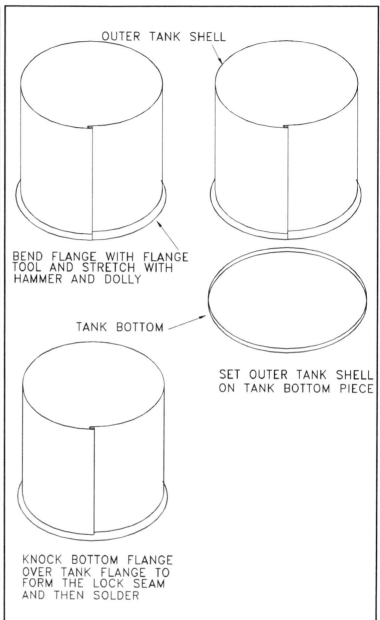

OUTER TANK SHELL

BEND FLANGE WITH FLANGE
TOOL AND STRETCH WITH
HAMMER AND DOLLY

TANK BOTTOM

SET OUTER TANK SHELL
ON TANK BOTTOM PIECE

KNOCK BOTTOM FLANGE
OVER TANK FLANGE TO
FORM THE LOCK SEAM
AND THEN SOLDER

FIGURE 31 Attaching the outer tank bottom to the cylinder

bottom flange over the cylinder flange to form the lock seam. Coat the seam with flux and solder. Fill the tank with water and check for leaks. After repairing any leaks that may exist drill a 1/2" hole for the drain line in the side of the tank. Locate the hole 10-1/2" from the top edge of the tank and opposite the seam side. Swedge the hole to 5/8", or until a piece of 1/2" copper water pipe fits snugly in it. See fig. 28. The hole should be swedged so that the flange is on the inside of the tank.

Building the outer tank; The outer tank cylinder is made from a sheet of 16 ounce copper that measures 54-1/4" x 18-3/4".

44

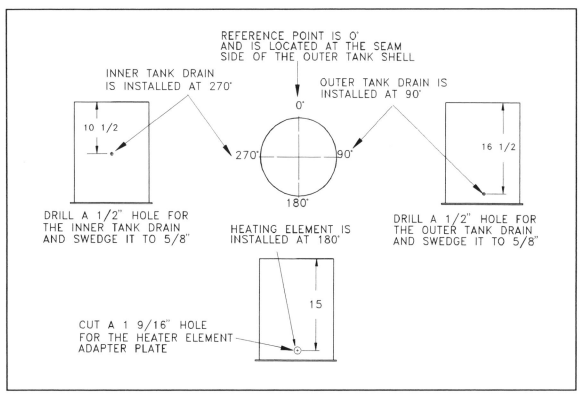

REFERENCE POINT IS 0°
AND IS LOCATED AT THE SEAM
SIDE OF THE OUTER TANK SHELL

INNER TANK DRAIN
IS INSTALLED AT 270°

OUTER TANK DRAIN IS
INSTALLED AT 90°

10 1/2

16 1/2

0°

270°   90°

180°

DRILL A 1/2" HOLE FOR
THE INNER TANK DRAIN
AND SWEDGE IT TO 5/8"

HEATING ELEMENT IS
INSTALLED AT 180°

DRILL A 1/2" HOLE FOR
THE OUTER TANK DRAIN
AND SWEDGE IT TO 5/8"

15

CUT A 1 9/16" HOLE
FOR THE HEATER ELEMENT
ADAPTER PLATE

FIGURE 32   **Locating and drilling the holes in the outer tank**

A line is scribed 3/8" from each end of the sheet showing the location of the flange bends.

The sheet is rolled to form a cylinder and the edges are hooked together, flattened and then soldered. See fig. 29. Bend a 3/8" flange on one end of the outer tank and stretch with a hammer and dolly.

The bottom of the outer tank is cut from a sheet of 16 ounce copper and has a radius of 9 1/8". A 3/8" flange is bent on its outer edge and then shrunk with a hammer and dolly. See fig. 30.

Assemble the outer tank and its bottom by placing the flange end of the outer tank cylinder on the tank bottom. Bend the bottom flange over the cylinder flange to form the lock seam and then solder it. See fig. 31. Put three holes in the outer tank. Two of the holes are drilled 1/2" and swedged to 5/8". The other hole has a diameter of 1-9/16" and is not swedged. One of the 5/8" swedged holes is for the inner tank drain and is located 10-1/2" from the top of the tank. The other 5/8" swedged hole is located 16-1/2" from the top of the tank and is for the outer tank drain. The 1-9/16" hole is for the heater element.

The exact degree location of these holes is not critical. The drawing in figure 32 shows the tank divided into four sections. the 0 degree mark is the reference point and is located at the seam side of the tank. The outer tank drain is located in a clockwise direction from 0 degrees at approximately 90 degrees. The hole for the heater element is located opposite the seam side of the tank or at approximately 180 degrees. The inner tank drain is located at 270 degrees.

The 1/2" holes can be made with a hole

45

saw or a drill bit. If you use a drill bit it is best to step drill the holes. Step drilling means starting a hole with a smaller size drill bit than is required and gradually increasing the size of the hole by increasing the drill sizes until the desired hole size is reached.

The 1-9/16" hole can be made with a hole saw or cut with a pair of tin snips. As mentioned earlier do not swedge the 1-9/16" hole.

Installing the heater element; Before the heater element can be installed a threaded adaptor plate must be mounted in the 1-9/16" hole located in the outer tank. Using this type of adapter plate with a screw in type heater element will make the job of changing elements in the future much easier.

Fig. 33 shows a drawing of the threaded adaptor plate and gasket used in the project. There are several brands of the adapter plate available and they can be found at most hardware stores. The adapter kit I used came from Ace Hardware and it's part # was 44101. There are a few other parts that come in the kit, but you only need the adapter plate and gasket.

Another item that is mounted to the tank with the adapter flange is the electrical box. See fig. 34. The box used is called a metal Project box and is available from Radio Shack. The box measures 5 1/4"L x 3"W x 2 1/8"D. It is installed for safety reasons and it covers the electrical connections made at the heater element. Four 3/8" mounting holes are drilled in the back of the box so that they line up with the holes in the adapter flange. Use the adaptor flange gasket as a pattern to mark the mounting hole locations. A 1-3/4" hole is cut in the back of the box so that it will fit over the element. This can be done with either a hole saw or tin snips. Two additional 3/4" holes

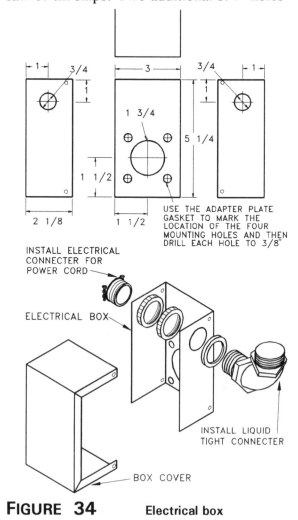

FIGURE 34   Electrical box

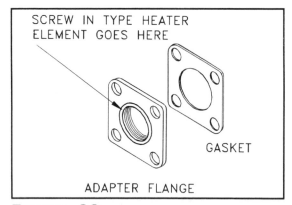

FIGURE 33   Heater element adapter flange and gasket

46

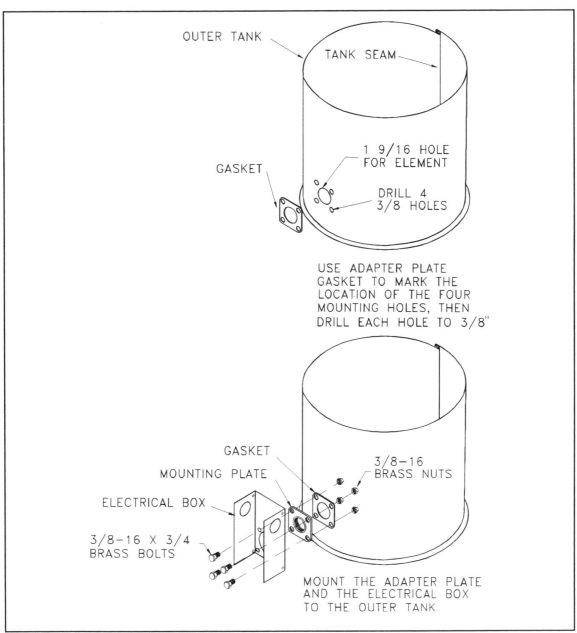

OUTER TANK

TANK SEAM

1 9/16 HOLE FOR ELEMENT

GASKET

DRILL 4 3/8 HOLES

USE ADAPTER PLATE GASKET TO MARK THE LOCATION OF THE FOUR MOUNTING HOLES, THEN DRILL EACH HOLE TO 3/8"

GASKET

MOUNTING PLATE

ELECTRICAL BOX

3/8-16 BRASS NUTS

3/8-16 X 3/4 BRASS BOLTS

MOUNT THE ADAPTER PLATE AND THE ELECTRICAL BOX TO THE OUTER TANK

FIGURE 35   Mounting the heater element adapter plate and electrical box.

are also drilled in the box. One 3/4" hole is drilled in the left side of the box and is for the power cord. The other 3/4" hole is drilled in the right side of the box and is for the wires which run through liquid tight conduit to the infinite range control switch that will be mounted on the column.

Line the adapter gasket up with the 1-9/16" hole in the outer tank. Mark the location of the four adapter plate mounting holes and then drill them to 3/8".

The adapter plate and electrical box are mounted with four 3/8-16 brass bolts and nuts. Brass fasteners are used because they won't rust. Fig. 35 shows the mounting procedure. To insure there won't be any

47

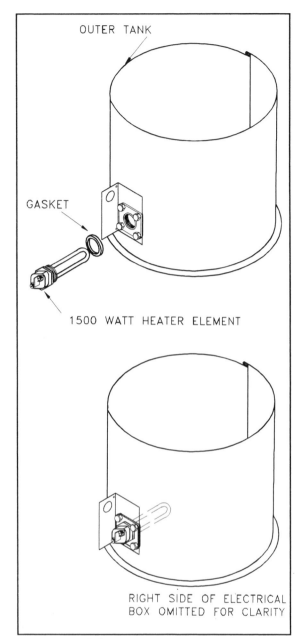

OUTER TANK

GASKET

1500 WATT HEATER ELEMENT

RIGHT SIDE OF ELECTRICAL
BOX OMITTED FOR CLARITY

FIGURE 36   Screwing in the
heating element.

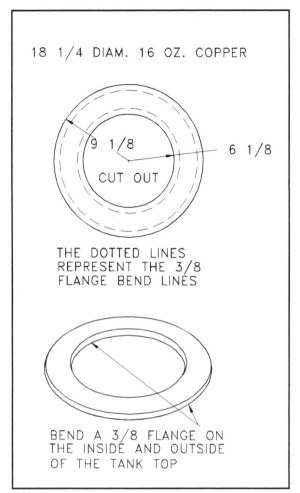

18 1/4 DIAM. 16 OZ. COPPER

9 1/8

6 1/8

CUT OUT

THE DOTTED LINES
REPRESENT THE 3/8
FLANGE BEND LINES

BEND A 3/8 FLANGE ON
THE INSIDE AND OUTSIDE
OF THE TANK TOP

FIGURE 37   Forming the top for
the outer tank

leaks it's a good idea to put a little silicon sealant on both sides of the gasket before mounting it.

The heater element and its gasket are installed next. It is a 1500 watt 120 volt element with a screw-in flange. There are several brands of this element available.

They can be found at most hardware stores. The element I used came from Ace Hardware and included the gasket. Its Ace part # was 44974. The element installation is simple. Place the rubber "O" ring gasket that comes in the package over the element on its thread side and screw the element into the tank. Tighten it using an element wrench until it is secure and the element is in a vertical position inside the tank. See fig. 36.

Putting the two tanks together; To make the outer tank top cut a circular piece

48

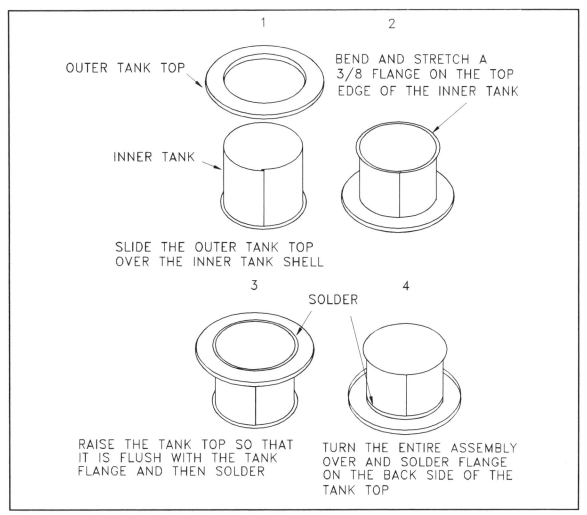

OUTER TANK TOP

BEND AND STRETCH A
3/8 FLANGE ON THE TOP
EDGE OF THE INNER TANK

INNER TANK

SLIDE THE OUTER TANK TOP
OVER THE INNER TANK SHELL

3
4

SOLDER

RAISE THE TANK TOP SO THAT
IT IS FLUSH WITH THE TANK
FLANGE AND THEN SOLDER

TURN THE ENTIRE ASSEMBLY
OVER AND SOLDER FLANGE
ON THE BACK SIDE OF THE
TANK TOP

FIGURE 38   Soldering the outer tank top to the top edge of the inner tank.

of 16 oz. copper to a diameter of 18-1/4". Scribe a radius mark 6-1/8" from the center. Cut the inside section out. See fig. 37. Bend and shrink a 3/8" flange on both the inside edge and on the outside edge.

To avoid confusion refer to the four steps shown in fig. 38 as we attach the outer tank top to the inner tank.

Position the tank top so that the flanges point down and slide it over the inner tank.

Then bend and stretch a 3/8" flange on the top edge of the inner tank.

Raise the tank cover up until it is flush with the tank flange. Clamp the seam together at opposite sides of the tank with "C" clamps. Apply flux to the seam area. Solder a short section of the seam close to one of the "C" clamps. After solder sets move "C" clamp over a little and solder another short section. Gradually work your way around the tank in this manner until the entire seam is soldered.

Turn the tank over and solder the seam on the back side as well, Being careful not to over heat it, because the solder on the other side could melt and cause the tank seam to separate.

The next step will be to put the two tanks

together by soldering the outer edge of the outer tank cover to the outer tank. Refer to fig. 39 as we begin.

Bend and stretch a 3/8" flange 90 degrees out on the top edge of the outer tank. Set the inner tank inside the outer tank. Line up the inner tank drain hole with the hole in the outer tank. The tank cover should fit nicely over the outer tank flange. Turn the tank assembly over being careful not to disturb the alignment of the drain holes. Knock the flange over with a hammer to form the seam. Coat the seam with flux and then solder it.

Drill two 3/4" holes in the top of the outer tank cover. Use a 3/4" hole saw or

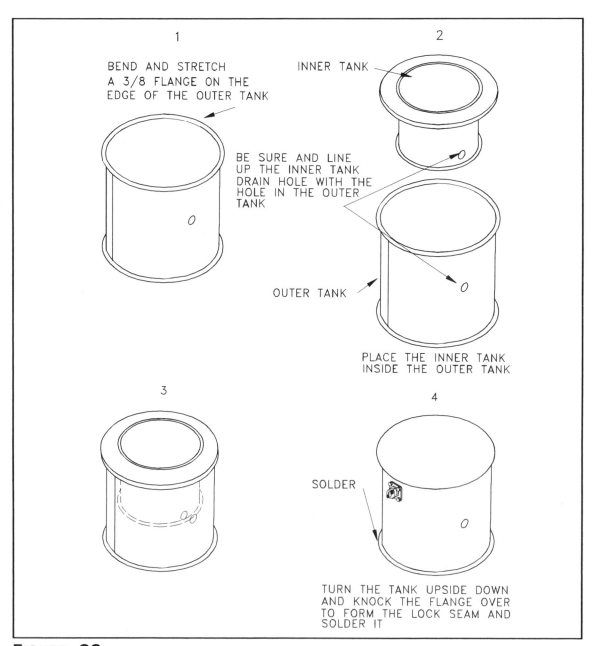

FIGURE 39   **Putting the two tanks together**

drill bit. Swedge the holes to 7/8". The location of the holes is not critical. Fig. 40 shows their location at 90 and 270 degrees with reference to the outer tank seam which is located at 0 degrees.

After the holes have been drilled and swedged insert a 3/4" CXM adapter in each hole and then solder them in place. CXM means copper on one end and male pipe thread on the other end.

Assemble and install the outer and inner tank drain lines. See fig. 41.

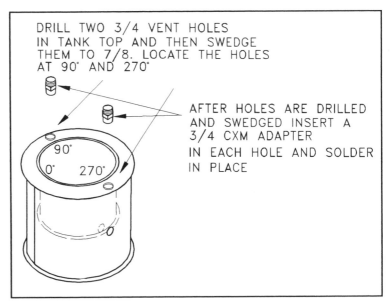

FIGURE 40   Installing the outer tank vents.

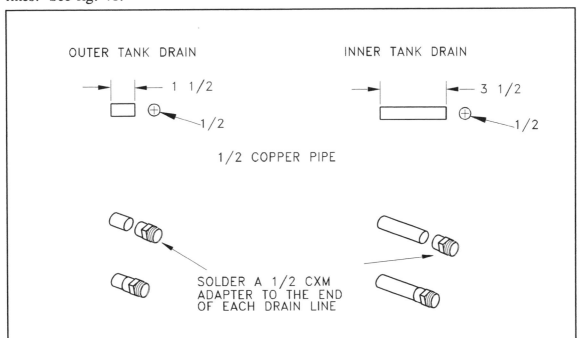

FIGURE 41   Making the inner and outer tank drains

The outer tank drain is a length of 1/2" copper water line 1-1/2" long with a 1/2" CXM adapter soldered to one end.

The inner tank drain is a length of 1/2" copper water line 3-1/2" long and it also has a 1/2" CXM adapter soldered to one end of it.

Put the outer drain line in the hole where it belongs and solder it in place. Insert the inner tank drain through its hole in the outer

51

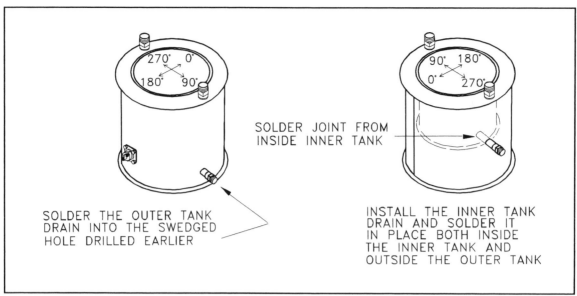

FIGURE 42   Installing the inner and outer tank drains.

tank and into its hole in the inner tank. Solder it in place inside the inner tank and also at the point where it enters the outer tank. See fig. 42.

Making the cone is very simple and you may refer to fig. 43 as we lay it out. The type of cone that we are going to build is called a truncated cone. This means that its top is cut off. In this case the top is cut off to make an opening for the rectifying column. The cone will also have a 3/8" flange formed on the bottom of it so that it can be soldered to the top of the inner tank.

Place a sheet of 16 ounce copper 26" long by 24" wide on a work bench. With a straight edge and scratch awl draw a full scale elevation drawing of the cone on the copper sheet. 3/4" has been added to the base width of the cone for the 3/8" flange. The cone elevation drawing is represented by points A, B, C in fig. 43. To allow plenty of room on the copper sheet for the layout, locate the base of the cone elevation drawing about 1/2" from bottom edge of the sheet and center it between the left and right edges.

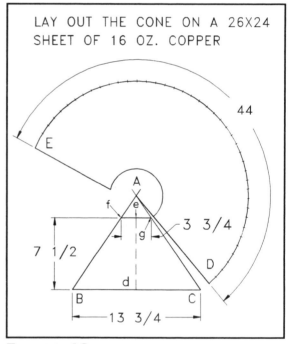

FIGURE 43   Cone layout.

To begin the elevation layout, scribe the 13-3/4" base line, point B,C. From the center of the base point d, scribe a perpendicular 7-1/2" line point d,e. Scribe a 3-1/4" line centered at point e. Line a straight edge up with point B

52

and f and scribe a line. Line the straight edge up with point C and g and scribe another line. The scribed lines B,f and C,g should intersect at point A. Point A will be the apex of our cone layout. This completes the elevation drawing of the cone.

The next step of the cone layout will be to scribe the arc D,E with point A being the center of the arc.

Measure the distance between points A, C. Set the dividers to that distance. If you don't have a set of dividers large enough to scribe the arc use the same method of scribing a circle that we discussed earlier. Match the distance between the nails on your homemade scribe tool with the distance between points A, C. Begin scribing the arc at point D.

We want the length of the arc to be the circumference of the cone plus a 3/8" seam allowance for each end. The circumference of the arc is found by multiplying the base diameter of 13-3/4 by 3.1416. This gives us a circumference of 43.2.

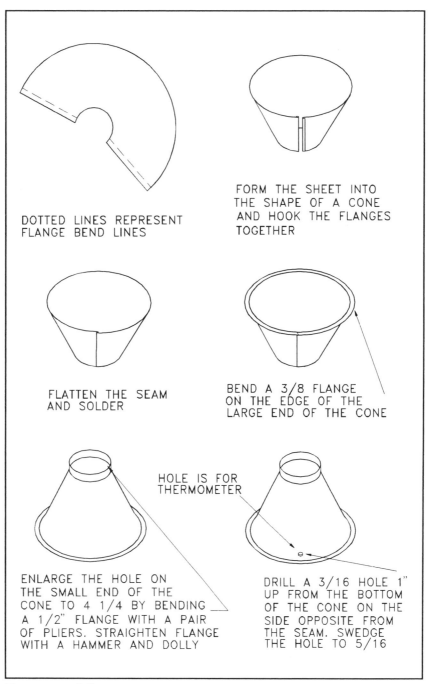

DOTTED LINES REPRESENT FLANGE BEND LINES

FORM THE SHEET INTO THE SHAPE OF A CONE AND HOOK THE FLANGES TOGETHER

FLATTEN THE SEAM AND SOLDER

BEND A 3/8 FLANGE ON THE EDGE OF THE LARGE END OF THE CONE

ENLARGE THE HOLE ON THE SMALL END OF THE CONE TO 4 1/4 BY BENDING A 1/2" FLANGE WITH A PAIR OF PLIERS. STRAIGHTEN FLANGE WITH A HAMMER AND DOLLY

HOLE IS FOR THERMOMETER

DRILL A 3/16 HOLE 1" UP FROM THE BOTTOM OF THE CONE ON THE SIDE OPPOSITE FROM THE SEAM. SWEDGE THE HOLE TO 5/16

FIGURE 44   **Forming the cone.**

Adding 3/4" for the seam allowance gives a final circumference of 43.95. We can round this off to 44". Lay out the circumference of the circle by setting a pair of dividers at 1" and stepping off the 44" distance beginning at point

53

D and continuing along the arc layout line.

Scribe a line from point D to point A and from A to point E. Set the dividers to match the distance between point A and g and scribe an arc. This is the truncated portion of the cone.

Cut the cone blank from the sheet and then refer to fig. 44 as we form the cone. Measure and mark each edge for a 3/8" flange. Bend the flange to form a hook on each end in opposing directions. Form the sheet into a cone and hook the flanges together. Flatten the seam and solder. Bend a 3/8" flange on the edge of the large end of the cone.

The small end of the cone is enlarged to 4-1/2" and has a 1/2" flange. Begin enlarging the hole by scribing a line 1/2" from the edge. This will be the flange bend line. Begin the flange bend with a pair of pliers and then straighten the flange with a hammer and dolly.

Drill a 3/16" hole 1" up from the edge of the large end of the cone and opposite the seam side. Swedge the hole to 5/16". The wash tank thermometer housing, will be located here.

Refer to fig. 45 as we make and install the wash tank thermometer housing. It is made

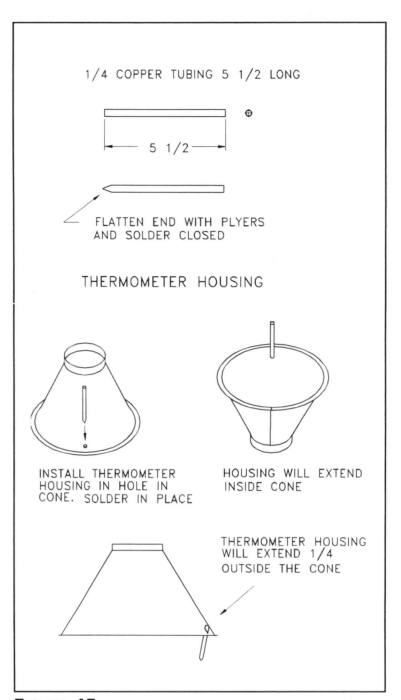

FIGURE 45 Installing the wash tank thermometer housing.

from a 5-1/2" long piece of 1/4" copper tubing. One end of the tubing is flattened and soldered.

The thermometer housing is installed in the 5/16" swedged hole located in the cone. It is placed in the hole at a slight angle and about

54

1/4" extends outside the cone, while the rest extends inside the cone. When it is positioned it can be soldered in place.

See fig. 46 as we attach the cone to the inner tank. The flange edge of the cone should line up nicely with the flange edge on top of the inner tank. Set the cone on top of the tank so that the thermometer housing is located on the same side as the outer tank drain and about 6" to the left of the outer tank vent.

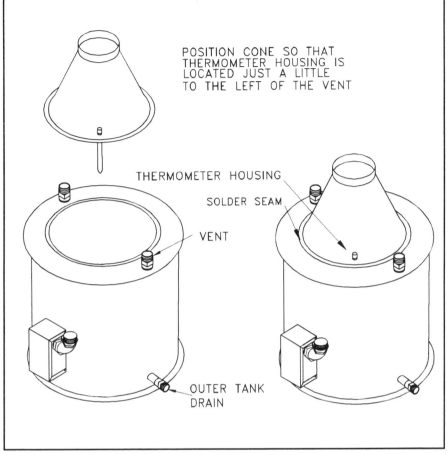

POSITION CONE SO THAT THERMOMETER HOUSING IS LOCATED JUST A LITTLE TO THE LEFT OF THE VENT

THERMOMETER HOUSING

SOLDER SEAM

VENT

OUTER TANK DRAIN

FIGURE 46   Soldering the cone to the inner tank

Since this will be a sweat joint make sure both flange surfaces are clean. Tin both surfaces before soldering. Set cone in place on top of the tank and solder.

The column extension is a 3" long 4-1/4" diameter cylinder that slips over and is soldered to the flange located on top of cone. See fig. 47. To make the column extension cut a piece of 16 ounce copper 14-1/2" long x 3" wide. Bend a 3/8" flange in opposite directions to form a hook on each end. Roll the sheet up to form a cylinder and hook the edges together to form a seam. Flatten and solder the seam.

The 18" x 4-1/4" rectifying column is made next and is shown in fig. 48. It is made from a sheet of 14-1/2" x 18", 16 ounce copper. Bend a 3/8" flange in opposite directions to form a hook on each end. Roll the sheet up to form a cylinder. Hook the edges together to form a seam. Flatten and solder the seam.

The rectifying column holds about twelve pounds of glass marbles. A disc with several 1/4" holes drilled in it and referred to as the column screen is installed in the bottom of the column to prevent the marbles from falling through. The holes are drilled in the disc so the vapors can enter the column and pass through the marbles. Fig. 49 shows the construction of the disc and flange.

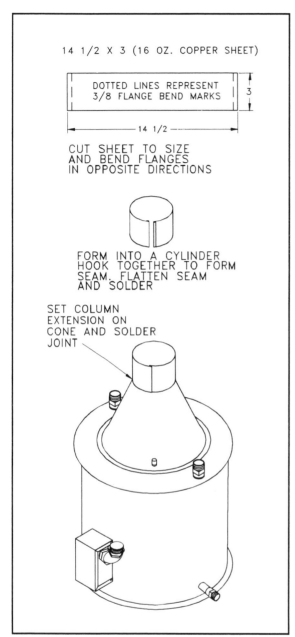

14 1/2 X 3 (16 OZ. COPPER SHEET)

DOTTED LINES REPRESENT
3/8 FLANGE BEND MARKS

3

14 1/2

CUT SHEET TO SIZE
AND BEND FLANGES
IN OPPOSITE DIRECTIONS

FORM INTO A CYLINDER
HOOK TOGETHER TO FORM
SEAM. FLATTEN SEAM
AND SOLDER

SET COLUMN
EXTENSION ON
CONE AND SOLDER
JOINT

FIGURE 47   The column extension.

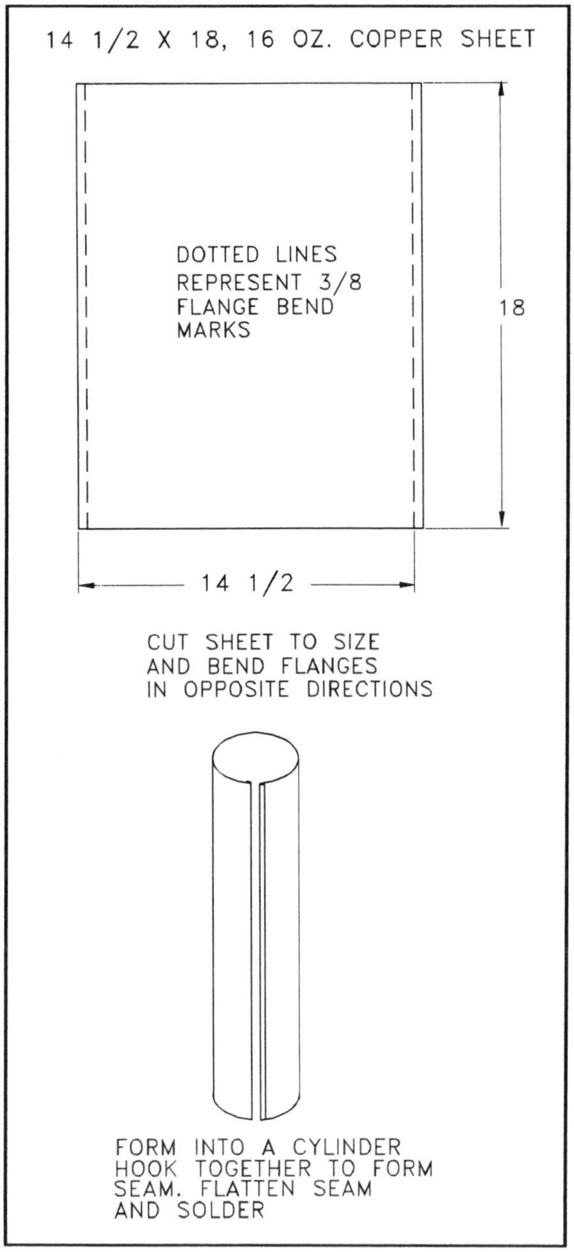

14 1/2 X 18, 16 OZ. COPPER SHEET

DOTTED LINES
REPRESENT 3/8
FLANGE BEND
MARKS

18

14 1/2

CUT SHEET TO SIZE
AND BEND FLANGES
IN OPPOSITE DIRECTIONS

FORM INTO A CYLINDER
HOOK TOGETHER TO FORM
SEAM. FLATTEN SEAM
AND SOLDER

FIGURE 48   The rectifying column.

Make the 4-1/8" diameter disc from 16 oz. copper and drill as many 1/4" holes in it as possible. Its hard to bend a flange on a disc this small so one is made from a 1" x13" piece of 16 oz. copper. Roll the copper up to form a ring. Butt the edges of the ring together and solder. Set the disc inside the ring and solder in place. Insert the flange side of the marble screen in one end of the rectifying column. Locate the bottom edge of the screen 2" from the end of the column. After the screen is installed it is

56

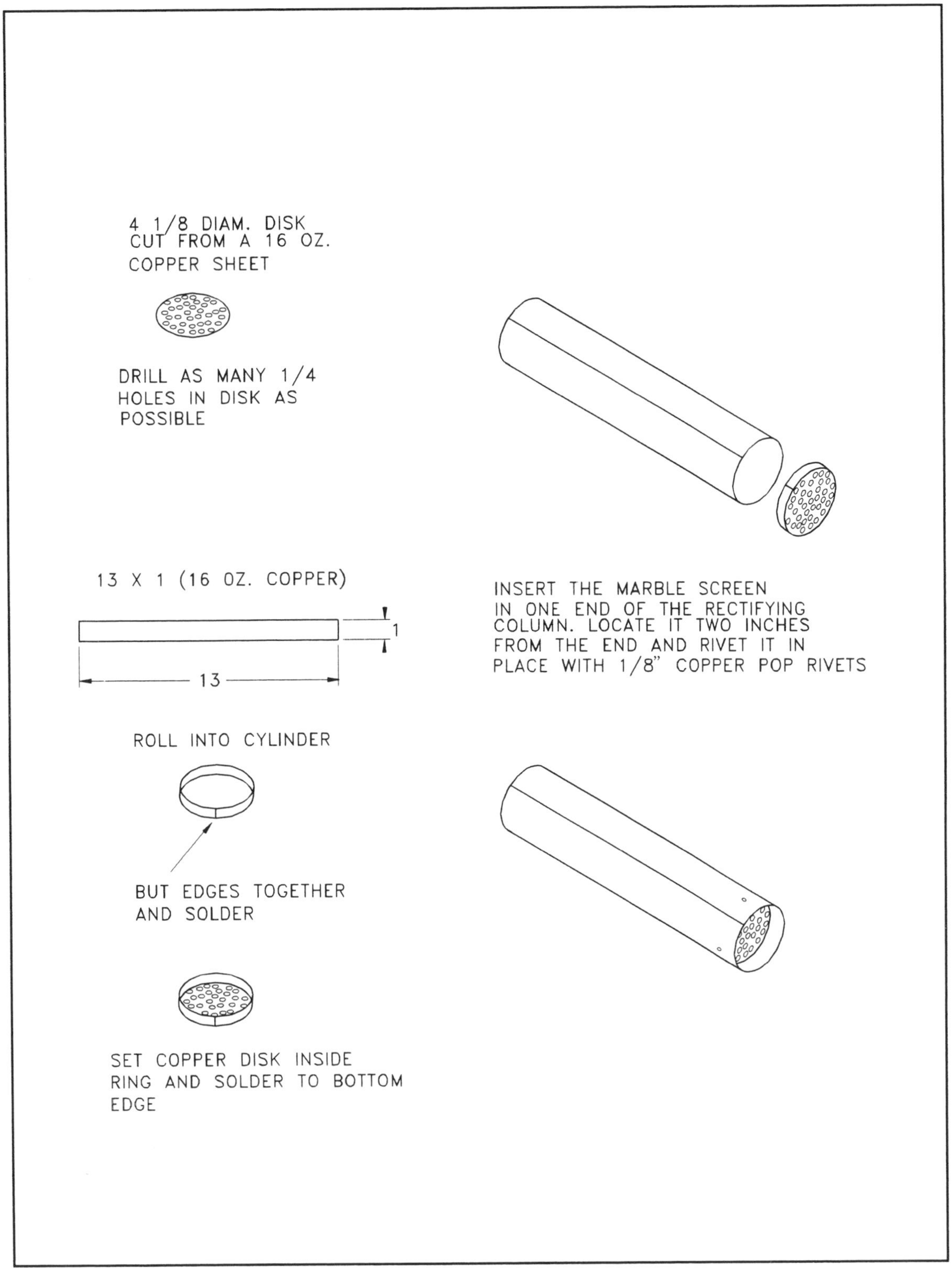

4 1/8 DIAM. DISK
CUT FROM A 16 OZ.
COPPER SHEET

DRILL AS MANY 1/4
HOLES IN DISK AS
POSSIBLE

13 X 1 (16 OZ. COPPER)

1

13

ROLL INTO CYLINDER

BUT EDGES TOGETHER
AND SOLDER

SET COPPER DISK INSIDE
RING AND SOLDER TO BOTTOM
EDGE

INSERT THE MARBLE SCREEN
IN ONE END OF THE RECTIFYING
COLUMN. LOCATE IT TWO INCHES
FROM THE END AND RIVET IT IN
PLACE WITH 1/8" COPPER POP RIVETS

FIGURE 49   **Making and installing the column screen.**

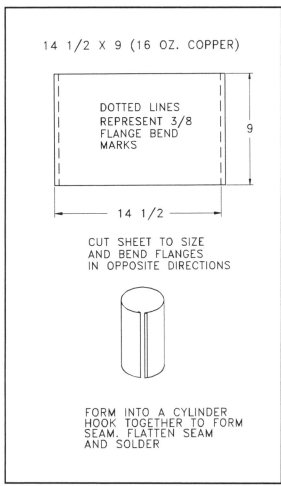

14 1/2 X 9 (16 OZ. COPPER)

DOTTED LINES
REPRESENT 3/8
FLANGE BEND
MARKS

9

14 1/2

CUT SHEET TO SIZE
AND BEND FLANGES
IN OPPOSITE DIRECTIONS

FORM INTO A CYLINDER
HOOK TOGETHER TO FORM
SEAM. FLATTEN SEAM
AND SOLDER

FIGURE 50   Making the upper column

secured with four 1/8" copper pop rivets spaced evenly around the outside of the column. See fig. 49.

Make the upper column next. It will be 9" long and have a diameter of 4-1/4". It is laid out on a sheet of 16 oz. copper that measures 14-1/2" x 9". See fig. 50. Bend a 3/8" flange in opposite directions on each end. Form the sheet into a cylinder and hook the edges together to form the seam. Flatten the seam and solder it.

The column top cone is made following the same guidelines used when making the cone for the inner tank. Since this cone is much smaller the layout dimensions are different. Also instead of bending the bottom

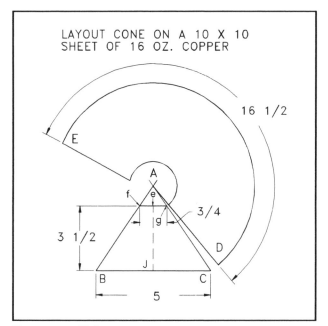

LAYOUT CONE ON A 10 X 10
SHEET OF 16 OZ. COPPER

16 1/2

E

A

f e

3/4

3 1/2

g

D

J

B   C

5

FIGURE 51 Laying out the column top cone

flange straight out, it is bent down so that the cone slips over the outside of the column. A thermometer housing is also installed in this cone. See figures 51, 52, and 53 as you layout and form the cone.

The next item of business will be to assemble the column. See fig. 54. To make the first connection slide a 4" no hub coupling over the cone extension. A 4" no hub coupling is made of rubber and is 2" long. It has an inside diameter of 4-1/4". There is an 1/8" rib located around the inside center section of the coupling. This rib will fit tightly against the edge of each column section being put together. This forms a tight seal that helps prevent leaks around the connections when the still is in operation. There is a piece of stainless steel screen or in some cases sheet metal and two hose type clamps that come attached to the outside of the coupling. The outer layer of stainless sheet or screen material adds support to the coupling and keeps the clamps from digging into the rubber as they are tightened. The no hub couplings are

58

available where ever plumbing supplies are sold.

Slide the rectifying column into the other end of the coupling. Tighten the hub clamps while Pushing down on the column to form a tight seal. Tighten the clamps securely being careful not to over tighten them. Over tightening would cause the column to cave in.

Before connecting the upper column to the rectifying column put approximately 12 pounds of glass marbles into the rectifying column. The glass marbles come in several different colors and are used in artificial flower arrangments. They can be purchased at most discount stores and are located in the hobby department or where the artificial flowers are sold.

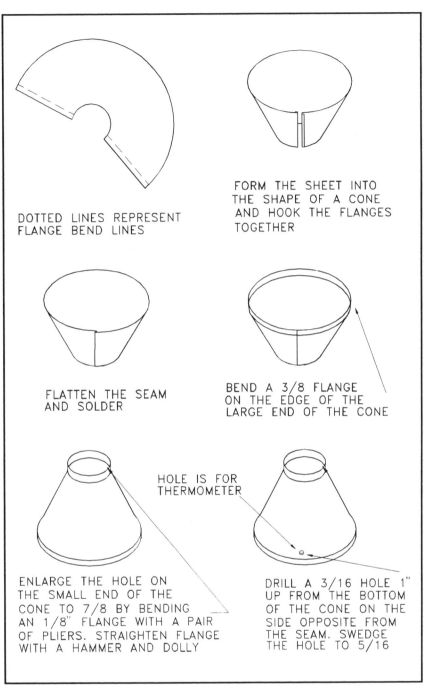

DOTTED LINES REPRESENT FLANGE BEND LINES

FORM THE SHEET INTO THE SHAPE OF A CONE AND HOOK THE FLANGES TOGETHER

FLATTEN THE SEAM AND SOLDER

BEND A 3/8 FLANGE ON THE EDGE OF THE LARGE END OF THE CONE

ENLARGE THE HOLE ON THE SMALL END OF THE CONE TO 7/8 BY BENDING AN 1/8" FLANGE WITH A PAIR OF PLIERS. STRAIGHTEN FLANGE WITH A HAMMER AND DOLLY

HOLE IS FOR THERMOMETER

DRILL A 3/16 HOLE 1" UP FROM THE BOTTOM OF THE CONE ON THE SIDE OPPOSITE FROM THE SEAM. SWEDGE THE HOLE TO 5/16

FIGURE 52   **Forming the upper column cone**

Slide a 4" no hub coupling over the top end of the rectifying column. Then put the upper column into the other end of the no hub coupling. Push down on the column to form the seal and tighten the coupling clamps.

Place the cone on top of the upper column so that the flange fits over the outside edge of the upper column and solder it in place.

The switch box mounting plate is made from a piece of copper or galvanized sheet metal that measures 8-7/8" x 9-3/4". See fig. 55 for the lay-out specifications. After making the necessary cuts the sheet is bent to form a box. The circular cut-outs on the top and bottom of the mounting plate are cut to fit the shape of the column. The mounting plate is placed on the column one inch from the top. It should be positioned so that it is directly above the heater element electrical box that is mounted on the outer tank. When in position the mounting plate can be soldered in place. See fig. 56.

Also see fig. 56 as we discuss the next couple of steps.

The faucets used for the inner and outer tank drains are standard laundry type with a 1/2" threaded female end. They can be installed at this time.

So that the condenser tank can be hooked to the column, cut a 1-1/2" long piece of 3/4" copper water pipe and place it in the swedged hole at the top of the column cone and solder it in place.

Install a 3/4" FHT x 3/4" FHT fitting to each of the outer tank vents. F.H.T. means female hose thread. The use of these fittings makes it easy to attach a garden hose for the purpose of filling the outer tank with water.

The infinite range switch electrical box is the same type of electrical box that we used to cover the heater element. The box can be made or you can purchase it from Radio Shack. Radio Shack calls the

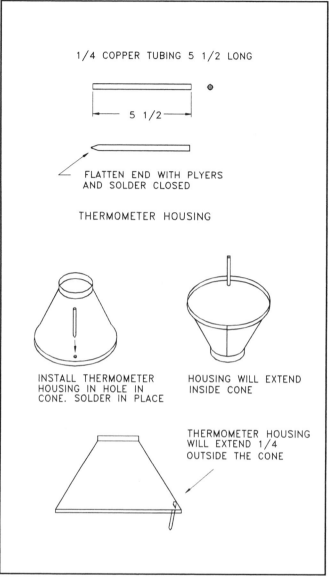

FIGURE 53   Installing the thermometer housing

box a metal project box and they come in several different sizes. The one were using measures 5 1/4L x 3W x 2 1/8 D. The box layout is shown in fig. 57. A 3/8" hole is drilled in the center of the front panel of the box cover for the switch shaft. Another 3/4" hole is drilled in the bottom of the box cover for the electrical conduit connecter. The conduit connecter can be installed at this time.

60

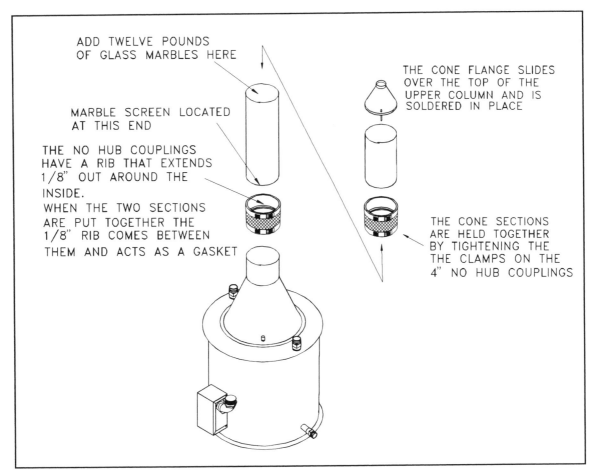

ADD TWELVE POUNDS
OF GLASS MARBLES HERE

MARBLE SCREEN LOCATED
AT THIS END

THE NO HUB COUPLINGS
HAVE A RIB THAT EXTENDS
1/8" OUT AROUND THE
INSIDE.
WHEN THE TWO SECTIONS
ARE PUT TOGETHER THE
1/8" RIB COMES BETWEEN
THEM AND ACTS AS A GASKET

THE CONE FLANGE SLIDES
OVER THE TOP OF THE
UPPER COLUMN AND IS
SOLDERED IN PLACE

THE CONE SECTIONS
ARE HELD TOGETHER
BY TIGHTENING THE
THE CLAMPS ON THE
4" NO HUB COUPLINGS

FIGURE 54   **Putting the column together**

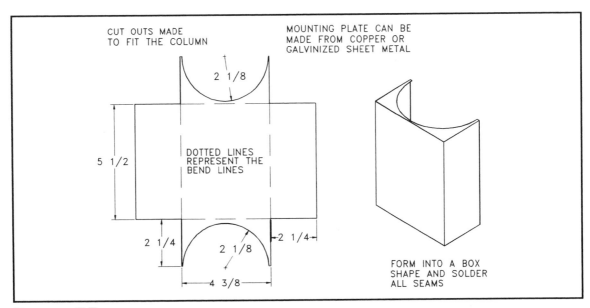

CUT OUTS MADE
TO FIT THE COLUMN

MOUNTING PLATE CAN BE
MADE FROM COPPER OR
GALVINIZED SHEET METAL

2 1/8

5 1/2

DOTTED LINES
REPRESENT THE
BEND LINES

2 1/4   2 1/4

2 1/8

4 3/8

FORM INTO A BOX
SHAPE AND SOLDER
ALL SEAMS

FIGURE 55 **Laying out the switch electrical box mounting plate**

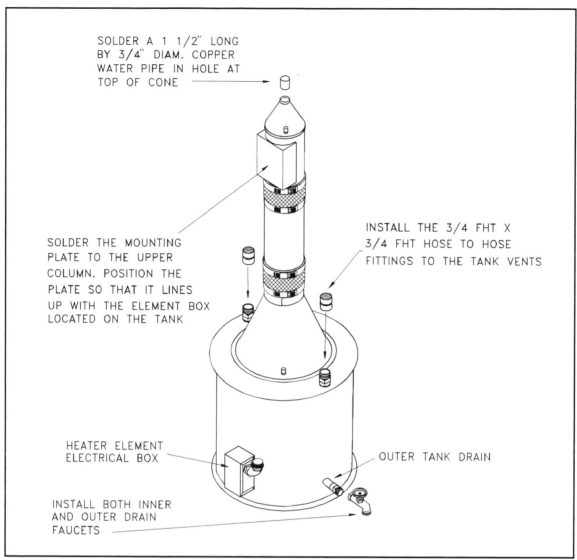

SOLDER A 1 1/2" LONG BY 3/4" DIAM. COPPER WATER PIPE IN HOLE AT TOP OF CONE

SOLDER THE MOUNTING PLATE TO THE UPPER COLUMN. POSITION THE PLATE SO THAT IT LINES UP WITH THE ELEMENT BOX LOCATED ON THE TANK

INSTALL THE 3/4 FHT X 3/4 FHT HOSE TO HOSE FITTINGS TO THE TANK VENTS

HEATER ELEMENT ELECTRICAL BOX

OUTER TANK DRAIN

INSTALL BOTH INNER AND OUTER DRAIN FAUCETS

FIGURE 56   **Installing the switch box mounting plate, hose fittings and drain faucets**

The type of electrical conduit used is called liquid tight. The size used is 1/2". I used this type of conduit because it is flexible and very easy to cut and work with.

Fig. 57 shows a drawing of the infinite range switch used and how it is installed in the switch box cover. The switch I used is manufactured by the Robertshaw Company. The part number of the switch is 5500-135 and it is referred to by the company as a 5500 Infinite Range Switch Uni-Kit. It is rated at 15 amps at 120 volts resistive load.

I purchased my switch locally through a national supplier called Johnstone Supply. It can be purchased at most other wholesale appliance supply dealers. Similar switches are manufactured by other companies and are available under different part numbers. Ask your dealer.

Refer to fig. 58 as we attach the switch box to the column and run the necessary wires.

Begin by centering the infinite range switch electrical box on its mounting plate

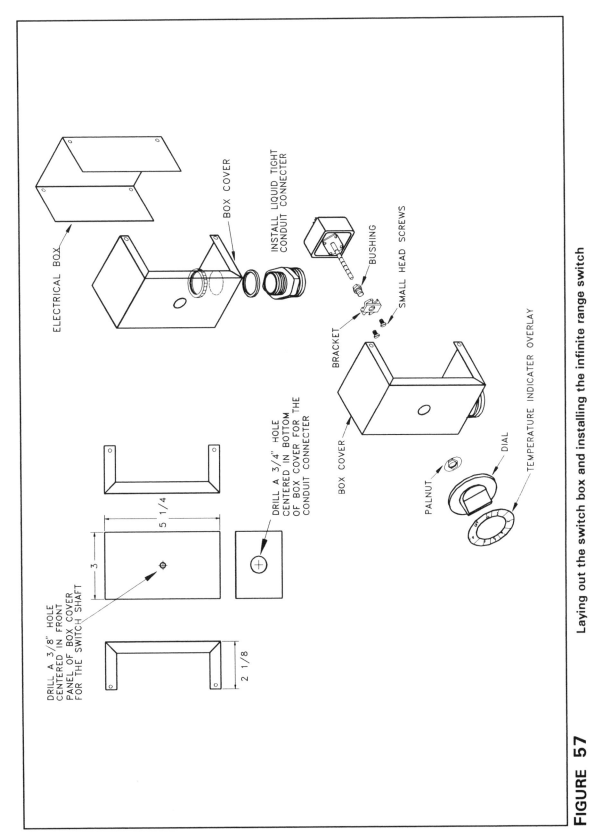

ELECTRICAL BOX

BOX COVER

INSTALL LIQUID TIGHT
CONDUIT CONNECTER

BUSHING

SMALL HEAD SCREWS

BRACKET

BOX COVER

PALNUT

DIAL

TEMPERATURE INDICATER OVERLAY

DRILL A 3/8" HOLE
CENTERED IN FRONT
PANEL OF BOX COVER
FOR THE SWITCH SHAFT

5 1/4

3

2 1/8

DRILL A 3/4" HOLE
CENTERED IN BOTTOM
OF BOX COVER FOR THE
CONDUIT CONNECTER

FIGURE 57   Laying out the switch box and installing the infinite range switch

63

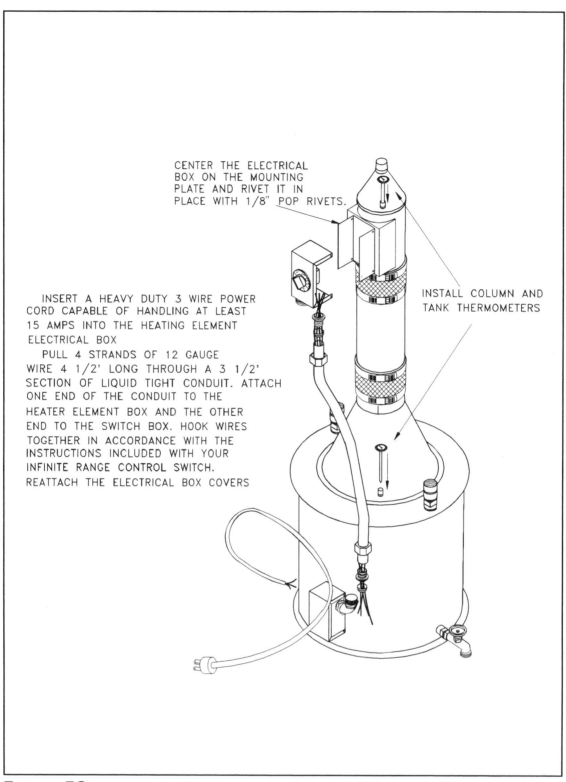

CENTER THE ELECTRICAL
BOX ON THE MOUNTING
PLATE AND RIVET IT IN
PLACE WITH 1/8" POP RIVETS.

INSTALL COLUMN AND
TANK THERMOMETERS

INSERT A HEAVY DUTY 3 WIRE POWER
CORD CAPABLE OF HANDLING AT LEAST
15 AMPS INTO THE HEATING ELEMENT
ELECTRICAL BOX
PULL 4 STRANDS OF 12 GAUGE
WIRE 4 1/2' LONG THROUGH A 3 1/2'
SECTION OF LIQUID TIGHT CONDUIT. ATTACH
ONE END OF THE CONDUIT TO THE
HEATER ELEMENT BOX AND THE OTHER
END TO THE SWITCH BOX. HOOK WIRES
TOGETHER IN ACCORDANCE WITH THE
INSTRUCTIONS INCLUDED WITH YOUR
INFINITE RANGE CONTROL SWITCH.
REATTACH THE ELECTRICAL BOX COVERS

FIGURE 58 Mounting the switch box, running the wires and installing the thermometers

located on the upper column. Secure the box to the mounting plate with four 1/8" pop rivets, one located at each corner of the box. Be careful not to drill through the column when drilling the 1/8" holes for the pop rivets.

Cut a piece of 1/2" liquid tight electrical conduit 3-1/2' long. As mentioned earlier

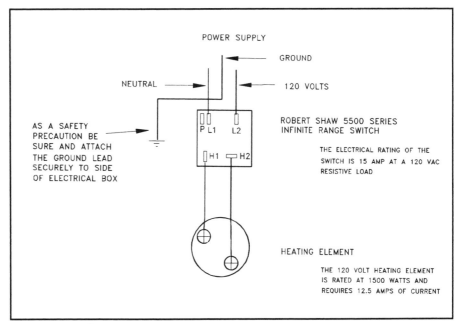

FIGURE 59   **Wiring diagram for a 15 amp 120 volt Robertshaw infinite range switch**

I used liquid tight conduit because it was inexpensive and easy to work with. If you have some other type of conduit on hand by all means use it.

Cut four pieces of 12 gauge copper wire 4-1/2' long. One of the wires should be black and one should be white. The other two wires should be a different color such as Red or brown. Pull them through the conduit so that about 6" of wire extends out each end. Attach the connecter nuts and gaskets to each end of the conduit. Remove the cover from the heater element electrical box. Thread the wires on one end of the conduit through the connecter in the heater element electrical box. Thread the wires on the other end of the conduit through the connecter on the switch box. Tighten the connecter nuts to secure the conduit.

Thread the wire end of a heavy duty 15 amp 3 wire power cord through the connecter located on the heater element box and securely tighten the wire clamp. The

cord should be the heavy duty type such as used on window air conditioners.

Before proceeding with the wiring portion of this project you should realize that there are many dangers involved in working with electricity and it is not possible for me to list them all. If you are not experienced working with electricity ask someone who is to help you. One of the main things to remember is *never work on an electrical circuit that is plugged in.* If you touch a live electrical wire you could receive an electrical shock. Electricity is especially dangerous around water so if you were to touch a live electrical wire while standing in a puddle of water the shock received could be fatal.

Another thing to remember is that all items powered by electricity should be grounded with a safety ground. Usually the safety ground is the green wire. In case of a short the safety ground will cause the electricity to run straight to ground.

Since the still is made of copper and is

filled with water, the risk of a fatal shock is great. I would recommend that the receptacle used be a GFI. GFI means ground fault interrupter. These receptacles are extremely sensitive to shorts in the circuit and when one is sensed they immediately shut off. They come in 15 and 20 amp sizes. The 1500 watt heating element used in the still uses 12.5 amps of current so the 15 amp size GFI should be sufficient.

Make sure all your wiring conforms to local codes and insurance regulations. The still should be hooked to a separate circuit and you should make sure that your electrical system will safely handle the required power.

Generally a detailed wiring diagram will be included with the purchase of your switch. Fig. 59 shows a wiring diagram which is only applicable for a 15 amp 120 volt Robertshaw 5500 series infinite range switch. If you use a different switch the wiring will be different, so always check the instructions that come with the electrical parts you purchase before attempting to hook them up.

A few basics involved in AC wiring are that the white wire is usually the neutral wire which means that there should be no voltage between it and ground. The black wire is usually the hot wire and there should be line voltage between it and ground. The green wire is usually the safety ground and there should be no voltage between it and ground. In working with electricity it is always a good idea to check the wires for voltage with a meter before you touch them. You never know when the person before you may have wired something backwards. I have come across these types of situations before and the results can be shocking.

Begin the wiring process at the heater element electrical box. The green wire located in the power cord should first be attached with a screw securely to the side of the electrical box. Match the black wire in the power cord with the black wire in the conduit. Twist them together and secure with a wire nut. Match the white wire from the power cord with the white wire in the conduit. Twist them together and secure them with a wire nut. Hook one red wire to each terminal on the heater element. Replace the cover on the electrical box.

Move to the infinite range switch box. Solder a female .250 quick disconnect terminal to the end of each wire. The black wire which carries 120 volts hooks to L2. The white or neutral wire hooks to L1. The red wires that go to the heating element attach to H1 and H2. Remember always check the instructions that come with the electrical components you buy before you begin any wiring. When you are through wiring replace the electrical box covers.

IMPORTANT! MAKE SURE THE WATER TANK HAS BEEN FILLED BEFORE YOU TEST THE HEATING ELEMENT. If tested in a dry tank the element will be ruined almost immediately. To fill the outer tank with water simply hook a garden hose to one of the fittings located on the outer tank vents and turn the water on. You can watch the water level rise by looking in the other vent hole with a flashlight. Be careful not to over fill the tank and get water in your electrical box. Its a good idea to have a shut off valve located between the water hose and the tank so that you can immediately turn the water off when it reaches the desired level. When the tank is full disconnect the water hose and clean up any water that may have spilled.

Make sure the infinite range switch is in the off position and then plug in the still. Turn the range switch on. If everything is working properly you will begin to hear the sound of the element heating the water almost immediately.

The still portion of the project is complete. Fig. 60 shows a drawing of the completed still.

Building the condenser;

The final part of the construction project will be building the condenser. It consists of a 20' length of 1/2" type L (heavy wall) copper tubing formed into a 7-1/2" diameter coil that is approximately 3 feet long. The coil is installed inside a 42" length of 8" diameter galvanized furnace pipe which serves as the condenser tank. A top and bottom is installed on the tank. The alcohol vapors are cooled by water that enters through a garden hose that is connected to a fitting located at the bottom of the tank and exits through another garden hose that is attached to a fitting located at the top of the tank.

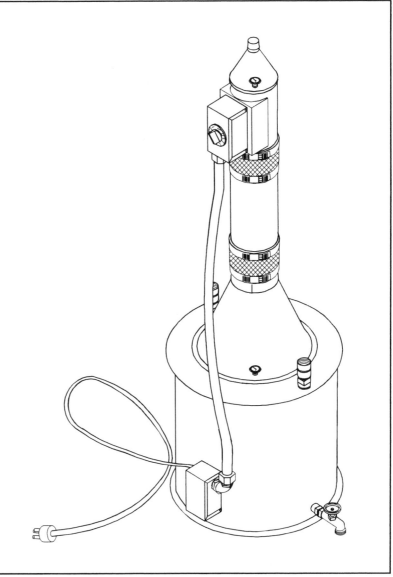

FIGURE 60   **The still complete**

The copper coil is made first. See fig. 61. The best way to form the tubing into a coil is to wrap it around a post, or a plastic or metal pipe that has a six inch diameter and is 4 or 5 feet long. Tape one end of the tubing in place with duct tape at the top of the 6" post and simply bend the tubing around the post. Bend it gradually and carefully so that it won't kink. Soft copper naturally hardens when it is bent so try to bend it where you want it the first time, because it will be much harder to bend the second time. Space each coil about 3" apart and make sure the entire coil runs down hill. Any high places in the coil will cause it to clog up later. Let about 10" of coil run straight up at the top

67

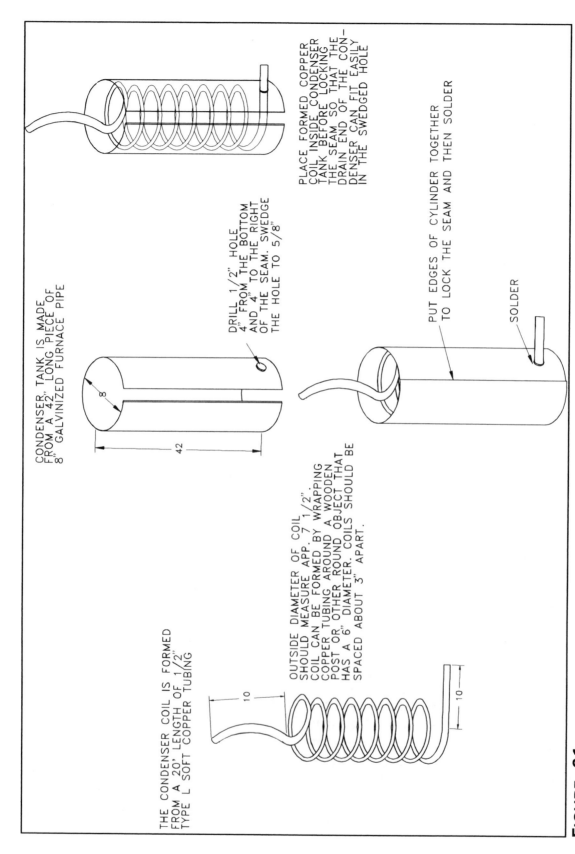

PLACE FORMED COPPER COIL INSIDE CONDENSER TANK BEFORE LOCKING THE SEAM SO THAT THE DRAIN END OF THE CONDENSER CAN FIT EASILY IN THE SWEDGED HOLE

CONDENSER TANK IS MADE FROM A 42" LONG PIECE OF 8" GALVINIZED FURNACE PIPE

DRILL 1/2" HOLE 4" FROM THE BOTTOM AND 4" TO THE RIGHT OF THE SEAM. SWEDGE THE HOLE TO 5/8"

PUT EDGES OF CYLINDER TOGETHER TO LOCK THE SEAM AND THEN SOLDER

SOLDER

THE CONDENSER COIL IS FORMED FROM A 20' LENGTH OF 1/2" TYPE L SOFT COPPER TUBING

OUTSIDE DIAMETER OF COIL SHOULD MEASURE APP. 7 1/2". COIL CAN BE FORMED BY WRAPPING COPPER TUBING AROUND A WOODEN POST OR OTHER ROUND OBJECT THAT HAS A 6" DIAMETER. COILS SHOULD BE SPACED ABOUT 3" APART.

FIGURE 61   Forming the condenser coil and the condenser tank and installing the coil in the tank

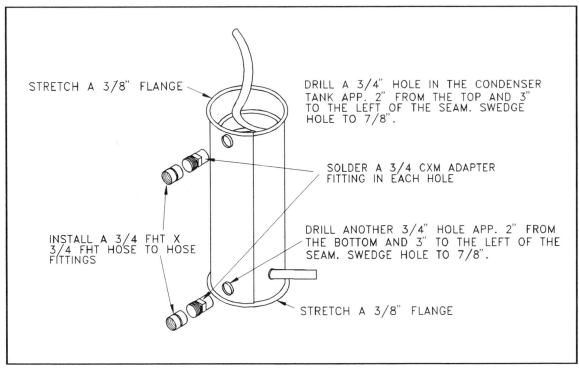

STRETCH A 3/8" FLANGE

DRILL A 3/4" HOLE IN THE CONDENSER TANK APP. 2" FROM THE TOP AND 3" TO THE LEFT OF THE SEAM. SWEDGE HOLE TO 7/8".

SOLDER A 3/4 CXM ADAPTER FITTING IN EACH HOLE

INSTALL A 3/4 FHT X 3/4 FHT HOSE TO HOSE FITTINGS

DRILL ANOTHER 3/4" HOLE APP. 2" FROM THE BOTTOM AND 3" TO THE LEFT OF THE SEAM. SWEDGE HOLE TO 7/8".

STRETCH A 3/8" FLANGE

FIGURE 62   Installing the water cooling inlet and outlet

and let 10" run straight out at the bottom as shown in fig. 61.

The 8" diameter furnace pipe used for the condenser tank comes in 5' lengths. Trim the pipe to 42". The left over pipe can be used for making the top and bottom pieces for the tank. Before putting the pipe seam together drill a 1/2" hole 4" from the bottom and 4" to the right of the seam edge at the bottom of the tank. Swedge the hole to 5/8". The coil exits through this hole. Place the formed coil inside the condenser tank before locking the seam. Fit the drain end of the coil through the swedged hole. Lock the seam edges of the pipe together. Solder both the tank seam and the flange around the drain end of the coil.

Drill two 3/4" holes in the tank. Locate the water outlet hole 2" from the top of the tank and 3" to the left of the seam. Locate the water inlet hole 2" from the bottom of the tank and 3" to the left of the seam.

Swedge each hole to 7/8". Solder a 3/4" CXM adaptor in each hole. Screw a 3/4" FHT x 3/4" FHT fitting to each adapter. See fig. 62. *(Special note; The water cooling inlet and outlet for the condenser tank can also be located on the back side of the condenser depending on where your water source is located.)*

Stretch a 3/8" flange on each end of the condenser tank. See fig. 62. Cut two 8-3/4" diameter sheet metal discs from the left over furnace pipe material. Raise a 3/8" flange around the edge of each piece. See fig. 63.

Set the condenser tank on one of the flanged discs. Bend the disc flange over the tank flange. Flatten and solder the seam. Align the other flanged disc up with the top of the tank. Since the exact location of the hole needed in the condenser tank top for the condenser coil line will vary you will have to mark its location, depending on your situation. When the location has been deter-

69

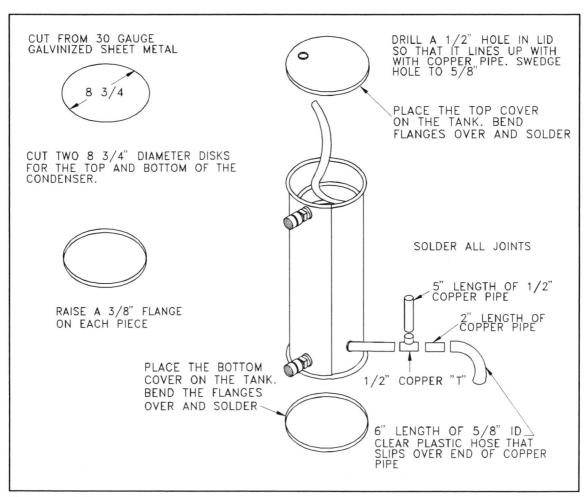

CUT FROM 30 GAUGE
GALVINIZED SHEET METAL

8 3/4

CUT TWO 8 3/4" DIAMETER DISKS
FOR THE TOP AND BOTTOM OF THE
CONDENSER.

RAISE A 3/8" FLANGE
ON EACH PIECE

PLACE THE BOTTOM
COVER ON THE TANK.
BEND THE FLANGES
OVER AND SOLDER

DRILL A 1/2" HOLE IN LID
SO THAT IT LINES UP WITH
WITH COPPER PIPE. SWEDGE
HOLE TO 5/8"

PLACE THE TOP COVER
ON THE TANK. BEND
FLANGES OVER AND SOLDER

SOLDER ALL JOINTS

5" LENGTH OF 1/2"
COPPER PIPE

2" LENGTH OF
COPPER PIPE

1/2" COPPER "T"

6" LENGTH OF 5/8" ID
CLEAR PLASTIC HOSE THAT
SLIPS OVER END OF COPPER
PIPE

FIGURE 63   Making and attaching the condenser tank top and bottom covers and
completing the condenser output line

mined drill the hole to 1/2" and then swedge it to 5/8". Set the top cover on the condenser, bend the flange over flat and solder the seam. Also solder the flange seam around the condenser line.

Solder a 1/2" copper "T" to the end of the copper tubing that exits the condenser. Cut a 5" length of 1/2" copper pipe and solder it to the top of the "T". This is the condenser vent and without it the alcohol would become vapor locked in the coil. Cut a 2" length of 1/2" copper pipe and solder it to the outlet side of the "T". Slide a 6" length of 5/8" clear plastic hose over the end of the condenser outlet. The condenser is

now complete and all that is left is to hook it up to the still.

Refer to fig. 64 as we begin hooking the condenser to the still. The condenser should be located one or two feet from the still. I located the condenser on the right side of the still, however it does not really matter which side you locate it on as long as it doesn't interfere with the operation of the control switch or your clear observation of the thermometers.

Begin by soldering a 3/4" 90 degree copper elbow to the 3/4" copper tube that extends out of the top of the column cone. Solder a 2" length of 3/4" copper pipe to the

70

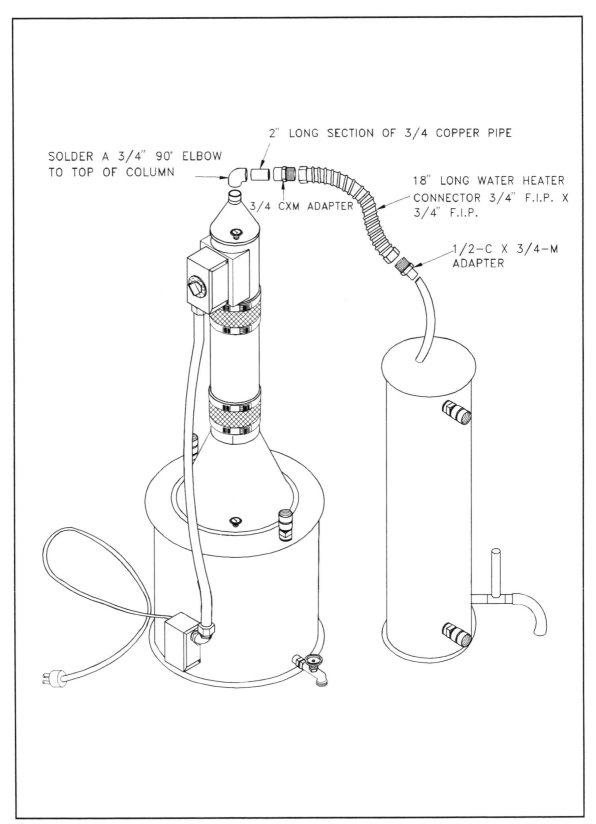

SOLDER A 3/4" 90° ELBOW
TO TOP OF COLUMN

2" LONG SECTION OF 3/4 COPPER PIPE

3/4 CXM ADAPTER

18" LONG WATER HEATER
CONNECTOR 3/4" F.I.P. X
3/4" F.I.P.

1/2-C X 3/4-M
ADAPTER

FIGURE 64   **Attaching the condenser tank to the still**

71

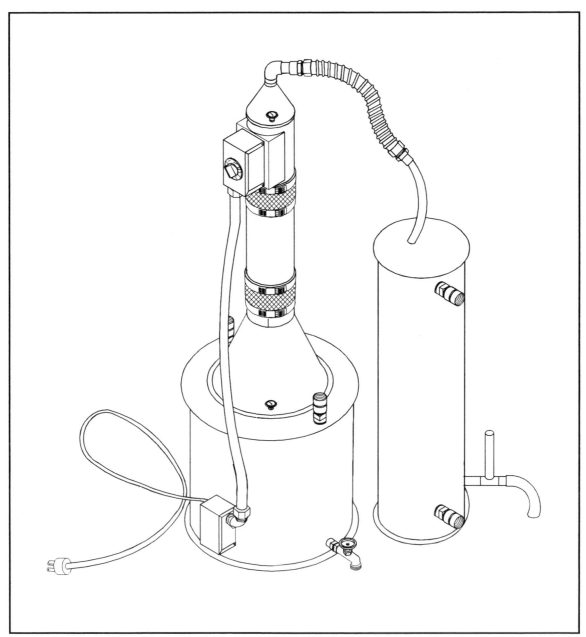

FIGURE 65   **The still complete**

end of the 90 degree elbow. Solder a 3/4 CxM adapter to the end of 3/4" copper pipe. Solder a 1/2 Copper x 3/4 Male adapter to the end of the condenser coil that sticks up out of the condenser tank. Connect the still to the condenser with a 3/4 F.I.P. x 3/4 F.I.P. flexible copper water heater connecter that is 18" long.

Congratulations you have just completed the construction of an alcohol producing still. Believe me, the hard part's over and the fun starts now because all that's left to do is produce alcohol with it. Before distilling alcohol with the still we need to review a few safety points. Always remember that alcohol and the fumes produced by it are extremely flammable. All distilling operations must take place outside and away from any type of flame or spark that could cause it to ignite. Whatever you do don't light or smoke a cigarette around the alcohol you produce. Another thing to remember is that the alcohol you will be producing will be almost pure at 170 to 190 proof. Strong stuff to say the least and for your own protection it's illegal to produce for drinking purposes. A single mouthful of pure alcohol could cause you to gasp for air and even stop breathing not to mention the risk of bodily damage caused by other harmful ingredients that may be in it. In this day and age it just does not pay to produce alcohol for human consumption. In most areas of the country high quality drinking stuff, much better than you could hope to make, is readily available legally and at a reasonable price at the liquor store. What I'm trying to say here is that there is no reason to break the law and that this project is meant only for the experimental

FIGURE 67 Filling the outer tank with water

production of grain alcohol for fuel use. When in operation the still should be constantly attended. This still can in no way be considered automatic and careful human attention must constantly be given to the operating temperatures of the wash tank and the column. If the temperatures in the column are allowed to approach 200 degrees it will cause a pressure build up inside the still which could result in a ruptured tank or explosion with hot stuff flying everywhere. It's your responsibility to use caution and good common sense in all aspects of this operation. BE CAREFUL!

Preparing the still for the first run;
Before you begin you should already have gone through the fermentation process and have 5 gallons of wash ready for distillation. Loosen the clamps on the no hub connecter

FIGURE 66   **Charging the still**

73

on the rectifying column located closest to the still tank. Remove the column in one piece and set it over to the side. This will leave a 4-1/4" opening to the inner tank. The wash will be added to the inner tank through this opening. Disconnect the condenser line from the still. See fig. 66. Also be sure both the inner and outer tank drain faucets are shut off before you begin filling the tanks. Now would also be a good time to set the still and condenser on a layer of concrete blocks. This raises the condenser off the ground so that a one gallon glass jug will fit under the coil outlet. It also allows easier access to the outer tank drain faucet.

Set a table up close to the still and set the 5 gallon glass carboy or bucket full of

FIGURE 69 Controlling the temperature with the infinite range switch

FIGURE 68   The still in operation

wash on the table. The bottom of the wash container must be at a level higher than the inner tank opening so that all of the wash can be siphoned into the still. Depending on the height of your table it may be necessary to set blocks under the wash container to raise it up. Place a siphon hose in the wash container and begin siphoning the wash into the inner tank. Remember you need to allow room for expansion caused by the heating process so do not put more than 5 gallons of wash in the inner tank.

While the wash is siphoning into the inner tank you can begin filling the outer tank with water. Hook a garden hose up to one of the vent fittings. You should have a shut off valve located on the garden hose at the vent fitting so the water can be quickly turned off when it reaches the desired level. It takes about 2-1/2 to 3 hours to heat cold water in the tank to 190 degrees.

You can speed the heating process up quite a bit if you add hot water to the outer tank instead of cold water. It's up to you. If you have access to a hot water faucet by all means use hot water. Another way to speed the heating process up is to wrap insulation around the tank.

Turn the water on and begin filling the outer tank. To allow room for expansion due to the heating process only fill the outer tank until it reaches a point 1" from the top.

When you're done siphoning the wash into the inner tank replace the column on the tank. Make sure the glass marbles are in the rectifying column. Tighten all the clamps on the no hub connecters and hook the condenser back up.

FIGURE 71   Collecting alcohol from the condenser

FIGURE 70 Checking the column temperature

When you're done filling the outer tank with water disconnect the hose from the tank. Under no circumstances leave the hose connected. To prevent a pressure build up and possible tank rupture or explosion both outer tank vents must be left open to allow steam to escape from the outer tank as the water is heated. Remember the still has been constructed of light weight material and no part of it can be considered a pressure vessel.

Make sure the inner tank and column thermometers are in place and that the infinite range switch is in the off position Plug the power cord into a GFI outlet. Remember never plug the still in when the infinite range switch is in the on position. Turn the infinite range switch on and adjust it to the high position.

Attach a garden hose to the inlet side of the condenser. A garden hose type shut off valve should also be installed between the garden hose and the condenser so the flow rate of the water can be easily adjusted. Attach another garden hose to the outlet side of the condenser and arrange it so that the water exiting the condenser will drain as far away from the still as possible. The reason for this is that we don't want any water accumulating around the still creating a shock hazard. Fill the condenser with water and then shut the water flow off for the time being. You will also need to place some kind of container under the condenser outlet to catch the alcohol when it begins to flow. A one gallon glass jug works well. Using a glass jug to catch the alcohol is nice because you're able to see the alcohol when it begins flowing. The heating process will take between 1-1/2 to 3 hours depending on whether you are starting with hot or cold water, or if you have wrapped the tank with insulation, so be patient while keeping a close eye on the wash tank and column temperatures.

The temperatures require careful attention and are fascinating to watch. As the still goes through the heating process there will be very little change in the column temperature until the inner tank reaches 190 to 195 degrees, at which time the column temperature will rise rapidly. As the column temperature begins to rise turn the water on so that it flows through the condenser at a steady but slow rate. Also be prepared to immediately adjust the temperature on the infinite range switch to LO as soon as the column temperature reaches 170 to 175 degrees. What we want to do is hold the temperature in the column as close to 173 degrees as possible. If you remember the vaporization point of alcohol is 173 degrees. As the column temperature drops adjust the range switch back up slightly to compensate. After a few adjustments the temperature will

begin to hold steady at 173 to 178 degrees in the column.

The hot alcohol vapor is now entering the condenser coil and is being rapidly cooled by the cold water circulating through the condenser tank. In order for the condenser to work properly it needs to cool the vapors to 60 degrees. It will take about five minutes for the alcohol vapor to make its way through the condenser and enter the receiver as a liquid.

Continue to watch the temperature of the wash and compare it with the column temperature. You will find that in order to maintain the proper column temperature, the range switch will have to be adjusted upward from time to time causing the temperature in the wash tank to gradually increase. As the alcohol evaporates the boiling temperature of the wash increases until it reaches 212 degrees. When the temperature in the wash tank reaches 212 degrees you can congratulate yourself on a successful operation because there will be no alcohol left in it and the distillation process will be complete.

Another interesting phenomenon is that when there is no alcohol in the wash tank the column temperature will begin dropping and it will be impossible to make it rise back up. When I tested my first still I put water in the wash tank and heated it up to 212 degrees and held it at that temperature for several hours. The upper column temperature never got above 90 degrees. I was concerned at first until I began to understand that the marbles caused the water vapors to continually condense back to a liquid. When vapor changes back to a liquid it gives up heat. Or in other words it cools down. This leads us once again to the reason why we are able to distill the alcohol from the wash . Because of the difference in the boiling temperatures of alcohol and water the marbles can be at 173 degrees and cause water vapor to condense on them while at the same time cause alcohol vapor to evaporate from them. The water drops back into the tank and the alcohol vapor rises up to the top of the column. In simple terms the still will not work unless the wash contains alcohol.

When the distilation is complete I am sure you will be curious as to what proof alcohol you have produced. The best way to get a proof reading of your alcohol is by the use of an alcohol hydrometer with a Proof and Traile's Scale on it. This instrument is similar in appearance to the sugar scale hydrometer we used to check the potential alcohol of our wort, except this one measures the percentage of alcohol and the proof of distilled spirits. You can purchase

an alcohol hydrometer with instructions for use, as well as many other related items from E.C. Kraus, 9001 East 24 Highway, P.O. Box 7850, Independence, Missouri 64053 or you can call them at 1-816-254-7448.

When the distilling operation is complete drain the inner tank and disassemble the column so that it and the tank can be cleaned. The still should be cleaned with a mild bleach and water solution and rinsed well before and after each use. The exterior copper surfaces can be kept clean and shiny with the use of a good quality copper cleaner.

Be sure and keep good records of each production run in case you are audited by the ATF. They would want to know production dates, how much produced, and what you did with what you produced.

Alcohol as a motor fuel;

Before we begin a discussion on alcohol as a motor fuel, I want to advise you not to use it in your automobile unless you check with the manufacturer. Today's modern engines with their computer designed fuel injection and pollution control devices are set up and designed to burn gasoline. These engines could be seriously damaged by using any other fuel than that which they were designed to burn.

An entire book could be written on the use of alcohol as a motor fuel. My knowledge only extends to my own personal experience and to a few things that I have read about, but have not had an opportunity to try. It would be impossible for me to detail the alcohol conversion process for every engine and I will not make an attempt

to do so. I will however try to relay some of the basics to you, but the rest is up to you. There is certainly room for a great deal of study on this subject and I would encourage you to do so before you begin experimentation. It would also be a good idea to experiment on an old engine or vehicle that you do not care about. That way if something goes wrong you won't experience a big loss.

Make sure that your still is producing alcohol of a consistent proof reading before you use it in your fuel experiments. The reason for this is that different carburetor adjustments would be needed for each varying percentage of purity. For instance if your fuel is 20% water the orifice of the carburetor would have to be larger than if your fuel was 5% water. Some tolerance is okay, but an extreme variance of 10% or more would require different adjustments.

Generally we live in a world of gasoline powered engines. To successfully burn alcohol in these engines requires a few changes. Most of these changes are done to the carburetor. Typically, engines and carburetors are different so it's best to have a book or manual that shows a complete drawing of your carburetor before you begin. You may also want to purchase a carburetor rebuilding kit so that any worn gaskets or other parts can be replaced if necessary.

Since burning alcohol requires less air and more fuel, the most important thing to be considered is enlarging the carburetor jets. The two jets to be concerned with in the average carburetor are the main metering jet and the idle jet. The main metering jet is the threaded brass plug with a hole drilled in it. It is usually located in the main well support or in the float body. The hole in the brass plug is called the main jet orifice. Its diameter dictates how rich or lean the air/fuel mixture will be when the engine is running at cruising speeds. The smaller the hole the less fuel will blend with the air and the leaner the mixture will be. The larger the hole the richer the mixture. Richer means more fuel less air. The original size of the main jet orifice will vary depending on the engine, carburetor size etc. Generally to convert an engine to alcohol it is necessary to enlarge the size of the main jet orifice by 20 to 40%. Since most of us are not engineers, we are unable to sit down with a calculator and determine the optimum size of the orifice. For that reason it is best to increase its size gradually. Reassemble the carburetor after each change and test the performance of the engine. This is called experimentation and it is painful and time consuming, but necessary.

Most carburetors will also require an idle mixture adjustment. The idle mixture jet is located at the base of the carburetor. It will have to be backed out a turn or two until the engine begins to run smooth.

Since alcohol has a higher octane rating than gasoline you will have to advance the timing considerably. As with everything else the timing adjustment will vary depending on your engine.

Ethanol burns a lot cooler than gasoline so you will no doubt have to replace your thermostat to one with a higher temperature rating. Another problem you may experience because of the cooler operating temperatures is a rough idle. Generally the motor will smooth out when it finally warms up.

If you are converting an engine with a fuel injection system it will be necessary to increase the size of the control jets 15 to 20%. I also understand that a fuel injection system does not require any special treatment for cold weather starting because the fuel is injected at pressures of 250 PSI. This causes the alcohol fuel to vaporize enough to ignite easily in the combustion chamber.

Engines running on a 100% alcohol mixture tend to perform better, but they are harder to start in cold weather. There are a couple of suggestions to remedy this problem that are worth mentioning. One method is to preheat the carburetor air before running it into the cylinders. This can be done electrically and only has to be done when starting the motor. Another method is to start the motor with gasoline and then switch over to alcohol after the engine is running.

Diesel engines can also be modified to run alcohol, but there are significant hurdles to overcome. Since they do not use conventional ignition systems it is difficult for alcohol to ignite in the combustion chamber. Also diesel injector pumps can not tolerate water. This poses a problem, especially since most alcohol contains at least some water. One thing for sure, diesel engines have closer tolerances and are much more costly to repair than gasoline engines. It might be a good idea to make sure you know what you are doing before you mess with them.

Since alcohol acts as a cleaning agent it will loosen a lot of particles in the fuel tank, lines and carburetor. This could cause the fuel filter to clog up. It will also clean the carbon, gum, and varnish from the engines internal parts causing your spark plugs to foul out.

I have an old 1949 Allis Chalmers tractor in which I have successfully used a 50-50 mixture of alcohol and gasoline without modifying the carburetor. It actually seems to run better. I have also used the same mixture in my lawn mower and chain saw.

Although I have not tried it I am told that a 100% mixture can be run in a 4-cycle lawn mower engine with only a couple of modifications. It would probably be best not to use your brand new $3,000.00 riding lawn mower in your alcohol experiments, so try the alcohol fuel in an older engine you don't care about first. Monitor its performance for a period of time and keep a record of the results. That way if something goes wrong there is no big loss. To begin the tests on your old engine, open the main metering jet (counterclockwise) about 1/2 turn. Then find the idle jet and open it between 1/2 and 3/4 turn. Fill the tank with alcohol and start the engine. Once the engine is started you can fine tune the carburetor adjustments. Use caution when operating

these engines on alcohol. Keep a close watch on the muffler and exhaust temperatures. If they seem excessively hot or the muffler starts to turn white, open the main jet slightly to enrich the mixture.

2-cycle engines such as chain saws and some industrial engines will also operate on 100% alcohol mixtures. You can modify them by adjusting the main and the idle jets as described previously for 4-cycle engines. It is also sometimes possible to buy special carburetors for these engines that are designed for alcohol use. These carburetors are adjustable and are often installed on engines used for racing, such as go-carts and motorcycles. One important point to remember when using alcohol in a 2-cycle engine is that you must use a synthetic oil in the fuel. Petroleum based oils do not mix well with alcohol. Some synthetic oil brands are Neo, Klotz, and Amsoil.

Alcohol can also be used as a home heating fuel and there are heaters designed for this purpose.

The idea of alcohol as a motor fuel is not a new one. It was first conceived over 70 years ago. Shortly before and during the depression several books, pamphlets, and university research papers had been written on the idea. In some countries dwindling fuel supplies were a problem as far back as the early 1930's. In 1931 the Philippines sold a blend of gasoline and alcohol called gasonol and special motors were designed to use it. In the early 30's Germany also mandated a 10 percent gasoline blend. And then again in the later years of World War II, because of Allied blockades and dwindling fuel supplies Germany was forced to build alcohol plants and develop plans to use ethanol exclusively.

As can be seen not all countries have been as fortunate as the United States. Until recent years we have had a seemingly endless supply of crude oil. The 1970's brought us to the realization that for our own protection and security it may be necessary to find other economical means of producing fuel. Sadly that realization was only temporary, because prices have stabilized and availability once again seems endless. What we don't realize is that we are no longer a self sufficient oil producing country and that we are importing 80 to 90 percent of all the oil we use. This is depleting our wealth because a great deal of our money is leaving this country to purchase the crude oil we use. This also makes us vulnerable to the demands and whims of the oil producing countries. Not to mention the fact that we are constantly having to deal with the constant turmoil and unrest that these countries seem to find themselves in.

I would say that it is only a matter of time before we find ourselves in another emergency situation frantically searching for alternate fuel sources. Alcohol may not be the best idea as a source of fuel, but it sure is something worth looking at.

Some of the reasons for alcohols slow development have been economical ones. Although gasoline and alcohol have to go through a similar distillation process to be produced there has been a cost difference in obtaining the crude product. To obtain crude oil all that is necessary is to drill a hole in the ground in the right spot, pump the crude out and it is ready for the distillery. To produce alcohol one must plant the grain and wait for it to grow and once the grain is harvested it must be allowed to ferment.

Over the last couple of years alcohol has become a much more competitive fuel to produce. The main reasons for this are that we can no longer drill the hole for the crude

oil in our own back yard. We've got to go half way around the world to do it. And then ship it back over to this country in huge tankers. Hardly a month goes by without hearing about one of these tankers running aground and spilling their load over the ocean. Remember the Exxon Valdiz in Alaska, and the environmental damage it caused. The clean up costs and the cost of the insurance that companies are forced to pay for these tankers also affect the cost of crude oil.

With the production of alcohol very little goes to waste. It is really an amazing example of recycling. When the sugar has been extracted from the grain it is resold as livestock feed. The carbon dioxide given off through fermentation is sold as dry ice or used to carbonate beverages. The only real expenditure is the fuel needed to heat the still. Some have even managed to bypass this expense by using solar energy to do the job.

Some interesting facts about alcohol;

Alcohol has a very low freezing point of 173 degrees below zero. For this reason alcohol was commonly used years ago as an antifreeze in car radiators.

Absolute alcohol is *Anhydrous* and *Hygroscopic*. Anhydrous means that it has all of its water removed. Hygroscopic means that it wants to absorb water. If you open a container of absolute alcohol it will soak up water from the humidity in the air. Automotive gas line antifreeze is actually alcohol and it works using this principle. If you suspect water in your gas line simply add the antifreeze (alcohol) to the gas tank. The alcohol will absorb any water that exists in the tank. It is impossible for your engine to burn pure water. But it will burn water when it is mixed with alcohol. Depending on how much water is in your tank the engine may cough and sputter but it will enable the engine to burn it through. Because of its low freezing point, alcohol when mixed with the water in your gas tank will also prevent it from freezing.

Ethanol alcohol has an octane rating of 106 using the common pump method. The octane rating of regular unleaded gas is 87. As you can see from the octane comparison, alcohol can be used on engines having a higher compression ratio than possible with regular unleaded gas.

Probably the most amazing and important property of ethanol is its ability to burn cleanly. It carries its own oxygen supply, so compared to petroleum fuel it only needs half as much oxygen for combustion. This is one of the reasons alcohol is used in rocket engines. Carrying less oxygen on a flight saves weight. In addition, the fuel burns cleanly easily and completely. The air/fuel ratio of gasoline is 15:1. The ratio for alcohol is 9:1

Gasoline releases many pollutants in the air as it burns, the worst being sulfur dioxide. The only primary exhaust products from a vehicle using ethanol are carbon dioxide and water. Another dangerous emission produced by gasoline engines is carbon monoxide which is harmful and sometimes fatal to both human and animal life. With ethanol fuel, carbon monoxide is not a serious issue. Only trace amounts of much less than 1% are emitted when burning alcohol. Typically 5 to 10% of the exhaust emissions of a gasoline engine are carbon monoxide. Several cities across the United States now mandate the use of Gasohol. (A 10 percent blend of ethanol with gasoline.) This has resulted in greatly reduced pollution levels in these cities and the requirement is sure to spread across the country.

Although ethanol is extremely dangerous to store and handle it is still much safer than gasoline. The flash point of gasoline is -45 degrees F. The flash point for ethanol is about +56 degrees F. This shows the wide safety margin of alcohol over gasoline. (The experts tell us that the flash point of a fuel is determined by heating the liquid until it ignites when a match is held close to its surface. The temperature of ignition is called the flash point.) Do not try to prove this fact. This procedure would not be safe and in fact would be down right dangerous.

Gasoline weighs 5.84 pounds per gallon. Ethanol weighs 6.60 pounds per gallon. The ignition temperature for gasoline is 720 degrees F. and 686 degrees F. for ethanol.

Material source list;

The Winemakers Market
4342 N. Essex Ave.
Springfield, MO 65803
Phone (417) 833-4145

They have a great mail order catalog that contains an excellent selection of premium winemaking and brewing products at reasonable prices. They were a big help to me and they have a good knowledge of all aspects of the brewing process. Yes they will custom grind your malt for you.

E.C. Kraus
P.O. Box 7850
Independence, MO 64053
Phone (816) 254-7448

Another great mail order company with a catalog containing an excellent selection of premium winemaking and brewing products at reasonable prices. You can purchase your Proof & Traile's scale alcohol hydrometer from them.

Bibliography;

The following books are available from:
Lindsay Publications,
P.O. Box 12,
Bradley, Ill 60915-0012
Phone (815) 933-3696

"Distillation of Alcohol and Denaturing".
By F.B. Wright.
An excellent book that contains lots of detailed information about the complete process of producing alcohol from beginning to end. It has an excellent section covering still and rectifying column design.

"Working Sheet Metal". By Dave Gingery.
Great book detailing sheet metal construction methods of all kinds. I used the methods described in this book to build the still.

"The Manufacture of Whiskey and Brandy Cordials."
A book originally published in 1937. It contains several intriguing methods, processes and layouts all geared for making hard liquor on a large commercial scale. It is not written for the home distiller. But it does have lots of information not normally available. The book also discusses several methods of still construction, malting, mashing and fermenting grains.

"Practical Distiller."
Another historical reprint from 1889 Shows methods from days gone by for distilling Brandies, Gin, Apple-jack, Rum, and other various essences and liqueurs.

"The New Complete Joy of Home Brewing". By Charlie Papazian.
Available from:
The Winemakers Market,
4332 N. Essx Ave.
Springfield, Mo 65803
This book deals with the art of brewing beer and it contains very good and complete descriptions of several methods of producing mash and on the fermenting process.